Simulation of assimilation, respiration and transpiration of crops

Simulation Monographs

Simulation Monographs is a series on computer simulation in agriculture and its supporting sciences

Simulation of assimilation, respiration and transpiration of crops

C. T. de Wit et al.

pudoc

Wageningen
Centre for Agricultural Publishing and Documentation
1978

Published by Pudoc outside the USA, Canada and Latin America

ISBN 90-220-0601-8

Printed in the Netherlands

Contents

The following research workers contributed towards the development of BACROS:

J. Goudriaan	Department of Theoretical Production Ecology
H. H. van Laar	Department of Theoretical Production Ecology
F. W. T. Penning de Vries	Department of Theoretical Production Ecology
R. Rabbinge	Department of Theoretical Production Ecology
H. van Keulen	Centre for Agrobiological Research (CABO)
W. Louwerse	Centre for Agrobiological Research (CABO)
L. Sibma	Centre for Agrobiological Research (CABO)
C. de Jonge	Computer centre

Ms. H. H. van Laar, R. Rabbinge and D. de Jonge edited the program and without their perseverance this publication would not exist. The typing by Ms. C. G. van Gulijk and the correction of the English by Ms. E. Brouns is kindly acknowledged.

1 Introduction

1.1 Purpose

For several years much attention has been paid to the construction of models for simulating crop growth and to their evaluation at the Department of Theoretical Production Ecology of the Agricultural University and at the Centre for Agrobiological Research (CABO), both in Wageningen. Although there have been several publications on the models and submodels treating various aspects of plant growth, this monograph is the first comprehensive report of the work. It has been written in such a way that the approach can be critically evaluated and further work on the subject may be stimulated in Wageningen and other centres.

A simulation program which covers all aspects of crop growth defeats its purpose. Such a program would be too large to be critically evaluated and to solve detailed problems that arise under field conditions. Therefore, the model described here is restricted to the potential growth situation, loosely defined as those growing conditions where the supply of water and nutrients is optimal for the crop, and there are no pests, diseases and weeds. Crop performance is then mainly affected by weather, crop husbandry and the properties of the plant species.

Simulation of plant growth in this situation is considered especially important because its results can be used as a reference for measurements in the field and thus sets goals and raises pertinent questions. Are yields lower than anticipated because some nutrients are not optimally supplied? Is there some disease which escapes attention? Are the results of the simulation over-optimistic and is more basic research needed?

Moreover, a simulation model of potential growth enables a quantitative evaluation of the influence of weather on yield and may provide a further framework for the analysis of weather and climate with respect to plant growth. What is then the influence of microclimate management through manipulation of the soil surface, or why does a crop like maize grow so well in temperate climates?

The growth of a crop may be divided roughly into three phases.

The germination and establishment, the vegetative growth and the storage phase in which seed setting, tuber formation or other similar storage processes occur. Until now, most attention has been paid to the growth in the vegetative stage for various reasons. Production of grassland is of primary importance under conditions in the Netherlands and is of increasing importance in other parts of the world. Since most grass is produced on permanent pastures, neither the germination and establishment phase, nor the seed-setting phase is very important. On the contrary, harvesting is generally done before the generative phase starts, to maintain quality of herbage and sward. Moreover the vegetative phase is of prime importance during the growth of species other than grass, which are also often grown for silage.

The purpose of the simulation and the present state of our knowledge are the main reasons for restricting attention to vegetative growth. This leads to a model that integrates knowledge of plant assimilation and carbon metabolism and of regulation of water flow through plants for the prediction of daily and seasonal patterns of crops in the 'potential growth situation', and evaluates the present concepts of growth mainly by comparing predicted and observed values and trends. The vegetative period in the 'potential growth situation' is the simplest case but still has much in common with actual field situations. Nevertheless its simulation requires a large number of processes and relations. Part of these are well known, but the model includes also relationships based on ad hoc assumptions.

In our opinion, simulation models, if they are to be useful at all, should form a bridge between reductionists, who analyse processes separated from their physical, chemical or biological background, and generalists who are interested in the performance of whole systems in which the individual processes operate in their natural context. Both the reductionist and the generalist should recognize their work in the simulation program. By comparing detailed output the generalist can independently evaluate how the model operates with field data, and the reductionist can determine whether the treatment of the processes that form the basis of the simulation model correspond with his ideas. To the reductionist simulation can be a guide to areas where research is most promising for further understanding of the system studied. To the generalist simulation extends his capability to envisage how a whole system functions.

It is possible to construct valuable models that simulate crop growth in many less optimal situations. Then it is necessary to describe primary production processes more simply, so that one can focus better on the major questions in the development of the model. Thus elaborately defined processes may be summarized and then incorporated in such broader models. This procedure has been used in other fields of research (van Keulen, 1975).

It is also possible to specialize further. Instead of being interested in the vegetative growth of crops, one may be interested only in the vegetative growth of one species: perennial rye grass, wheat, potatoes or maize. The present simulation was attempted on the assumption that the processes of vegetative growth of the main agricultural crops are similar because the underlying principles of plant physiology are the same. A change in the model from one plant species to another may thus be achieved by alteration of parameters only. These non-species-specific programs may work because the most complicated aspect of plant growth, the development of form and function, is only treated superficially at present.

1.2 Simulation technique

The simulation models used, are based on the assumption that the state of the system at any particular time can be expressed quantitatively and that changes in the system can be described mathematically. This assumption leads to the formulation of state determined models in which state variables, driving or forcing variables, rate variables, auxiliary variables and output variables can be distinguished.

A system may be defined as a limited part of reality, containing interacting elements. The totality of relations within a system is called the structure of the system; both systems and models have a structure. The physical limits of a system are obvious if the system is well isolated from its environment. Often, however, this is not so, and processes in one part of the system affect those in other parts. For example transpiration, assimilation and growth processes affect the composition of soil and air. On the field scale, however, these effects of a growing crop in the 'potential growth situation' are negligible, so that our model of primary production processes can be restricted to crop processes and weather within the crop.

State variables characterize and quantify all properties that describe the current state of the system. Examples of such variables are amount of biomass, number of animals, content of mineral elements in various parts of the system, amount of food, amount of poison, number of niches, water content, temperature of the soil and so on. In mathematical terms, state variables are quantified by the contents of integrals. Their values have to be known at the onset of simulation.

Driving or forcing variables are those that are not affected by processes within the system, but characterize the influence from outside. These may be macrometeorological variables, the amount of food added in course of time and so on. It should be realized that depending on the boundaries of the system to be simulated, the same variables may be classified either as state or driving variables.

Rate variables quantify the rates of change of the state variables. Their values are determined by the state variables and the driving variables according to rules formulated from knowledge of the underlying physiological, chemical and physical processes. These processes may be so complicated that the calculation process becomes much more lucid when use is made of properly chosen intermediate or auxiliary variables. Output variables are the quantities which the model produces for the user. Sometimes they are state variables, sometimes rates and sometimes auxiliary variables that may be calculated especially for the purpose.

In such state determined models, rates are not mutually dependent: each rate of change depends at any particular time on the values of the state and driving variables and can therefore be calculated independently of all other rates. Thus structural equations, that means *n* equations with *n* unknown rates, do not occur. The various sections of the model discussed interact nevertheless strongly, because the simulation is executed parallelly.

The time courses of the variables are generated in the model by calculating at an instant of time all rates, and realizing these over a short time interval, DELT. This procedure gives the value of the state variables at a time DELT later and it can then be repeated. The rectilinear or Euler method of integration is the most elementary one. It allows for discontinuous processes such as sudden leaf-fall, sudden cutting of crops or abrupt changes in weather, but forces the user to choose the time interval DELT small enough compared with the smallest time constant of the system. More sophisticated integration methods adapt the size of the time interval to the relative rates of change, but can only be used if all processes are continuous.

The use of the state determined system approach for the simulation of ecological processes has been analysed, discussed and illustrated by de Wit & Goudriaan (1974) in another monograph of this series. In this book, the simulation language CSMP S/360 (Continuous System Modelling Program, IBM manual SH20-0367-4) was used, both for programming and for the explicit formulation of structural relations. To understand the technical aspects of the present work it is useful to consult this monograph. The simulation programs presented here make use of CSMP III (IBM manual SH19-7001-2).

2 Outline of model and evaluation methods

A modelling effort, based on an analysis of the processes that are operative in a system, results in special models for various purposes. The models that are discussed here include the seasonal growth of closed crop surfaces and the daily course of assimilation, respiration, growth and transpiration. However, in other studies the same or similar model elements have been used to simulate the micro-weather in so far as it affects the growth and development of diseases, efficiency of water use under arid conditions, growth regulation in greenhouses or competitive phenomena. Therefore, the various processes are at first not discussed in the form in which they were incorporated in the main models, but in a somewhat broader way that enables expansion for special purposes. In the body of the model, the international unit system (SI) based on kg, m, s, °C and Joules as a derived heat unit is used, except for the unit of water vapour pressure (mbar), the unit of plant water stress (bar) and the unit of CO_2 concentration (volume parts per million, abbreviated as vppm).

2.1 Outline of model

First the Basic Crop Simulator (BACROS) is briefly described, to put in perspective the many detailed processes that are considered in the next chapters.

A crop in the vegetative phase of growth is considered, which is well supplied with water and nutrients. Growth of this crop is defined as increase in dry weight of the structural plant material, i.e. total dry weight exclusive of those organic substances that are classified as reserves. The model is based on physical, chemical and physiological processes, so that there is no restriction to the geographical range in which it can be applied. In Fig. 1 a simplified relational diagram of the simulation model is given; the rectangles represent state variables, the valves rates and the circles are auxiliary variables.

Micro-weather is calculated from the weather measured at screen height, the extinction of radiant energy from sun and sky within the

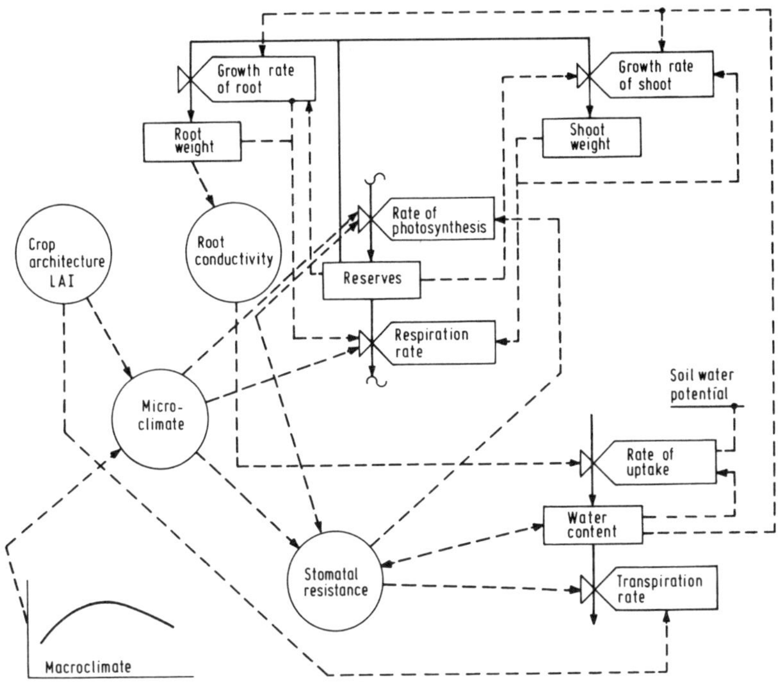

Fig. 1. Relational diagram of the simulation model. Rectangles represent state variables; Valves represent rates of change of the state variables; Circles represent intermediate or auxiliary variables or systems.

crop being taken into account. The infrared radiation from the canopy is also computed. A calculation of the distribution of radiation over the leaves is necessary for computation of assimilation and transpiration. The architecture of the crop determines this distribution of radiation and has to be defined. The extinction of turbulence in the canopy is also considered, so that transfer of heat, vapour and carbon dioxide can be computed. The ratio of latent and sensible heat exchange regulates to a large extent the micro-weather and this ratio is determined largely by stomatal behaviour. Basic models on heat transfer in the soil are available (de Wit & van Keulen, 1972). However, soil temperature is not simulated here in detail, it being assumed that this temperature follows the air temperature with a delay of 4 hours. Chapter 3 describes how the weather is modelled.

Readers interested in more detailed modelling of micro-weather are referred to Goudriaan (1977).

The assimilation of carbon dioxide by the canopy is calculated by adding the assimilation rates of the variously exposed leaves in successive leaf layers. These rates are dependent on light intensity, CO_2 concentration in the ambient air and resistance to CO_2 diffusion from the atmosphere towards the active sites. Transpiration and CO_2 assimilation interact strongly, not only because a relatively large transpiration may lead to loss of turgidity of the plant and subsequent closure of stomata, but also because a low rate of assimilation may lead to closure of stomata and low transpiration, through regulation of the CO_2 concentration in the stomatal cavity.

Respiration is the sum of maintenance respiration and growth respiration. The latter is caused by the conversion of reserves into structural material and is therefore proportional to the rate of growth. The intensity of growth respiration is affected by the chemical composition of the new material, which may be about equal to that of the plant. This intensity is independent of temperature, but growth respiration is indirectly influenced by temperature through the temperature dependence of the growth rate. Usually, CO_2 evolution resulting from translocation is included in the term for growth respiration. The rate of maintenance respiration depends on the turnover rates of proteins and the resynthesis of other degraded compounds and the maintenance of ionic gradients. This respiration process therefore depends largely on the chemical composition of the plant. The rate of maintenance respiration is sensitive to temperature.

The growth rate of the organs is dependent on the amount of reserves and temperature. Under internal water stress, growth of shoots is retarded by making a larger proportion of reserves available for growth of roots. By this mechanism a functional balance is maintained between root and shoot.

Up to now, no satisfactory solution has been found for the simulation of the growth of leaf surfaces in relation to the growth of leaf or shoot weight, so that these aspects are mimicked rather than simulated: their description is based on information obtained from field trials and not derived from knowledge of the underlying processes. Hence, the simulation programs become considerably limited in their application. However original research on plant physiology rather than model building is necessary to improve the situation here.

The water status of the plant is determined by the balance

between transpiration and water uptake from the soil. The transpiration rate of the crop is found by adding the transpiration rates of the variously exposed leaves in successive leaf layers of the crop. These rates are calculated from the radiation absorbed, resistance of the laminar layer, humidity and temperature of the ambient air and stomatal resistance. Stomatal resistance is either controlled through CO_2 concentration and assimilation or through the water status of the plant. The calculation also provides leaf temperatures, which are used in the photosynthesis section and averaged to give the crop temperature that affects growth and respiration. Water uptake is determined by the conductivity of the root system, the water status of the plant and that of the soil. The water status of the soil is assumed to be optimal ('field capacity'), so that the transport of water within the soil can be ignored. The conductivity of the root system is dependent on the amount of roots, their degree of suberization and soil temperature.

2.2 Experimental evaluation

2.2.1 Field trials

The most straightforward way to evaluate the results of simulation experiments is to compare them with the actual growth of crops. These comparisons have been made in various years and at different geographical locations for grasses, mainly perennial rye grass, a C_3 plant, maize, a C_4 plant, and wheat, a C_3 plant. Since morphogenesis is not simulated, it is necessary to introduce the leaf area index and the chemical composition of the crop in course of time as a forcing function. Hence deviations between simulated and observed rates of dry matter growth must be attributed solely to an unsatisfactory treatment of the main transfer processes such as assimilation, respiration and transpiration. However, actual growth rates are determined by periodic harvests and the variance of the difference between two successive yields is equal to the sum of the variances of these yields. Even in carefully executed experiments with several replicates, the variance of yield is at least $(200)^2$ $(\mathrm{kg\,ha^{-1}})^2$, so that the standard deviation of the difference is about $\sqrt{((200)^2+(200)^2)} =$ 280 $\mathrm{kg\,ha^{-1}}$. Since crop growth rates are often about 200 $\mathrm{kg\,ha^{-1}\,day^{-1}}$, time intervals of 14 days between harvests are needed to obtain an accuracy of 10 percent in the estimate of growth rate. Hence in this way only averages can be evaluated over rather long periods and an evaluation of day-by-day performance and performance throughout the day is an illusion.

2.2.2 Crop enclosure

Obviously another evaluation technique is needed, and partly for this reason, an installation was built that allows the continuous measurements of CO_2 assimilation and transpiration in crop enclosures in the field (Louwerse & Eikhoudt, 1975; Alberda *et al.*, 1977). With this installation, which is a modification of that of the Grassland Research Institute in Hurley (Leafe, 1972), the daily course of transfer processes can be investigated and the reactions of these processes to instantaneous modifications of the environment can be observed.

A crop area which may be $2 \times 2\ m^2$, is enclosed by a transparent cover, tightly placed on a frame, hammered into the soil to a depth of about 20 cm. The air in the enclosure is circulated over a cooling and heating unit at such a speed that it is thoroughly mixed in the enclosure. The air inside is kept at a slight overpressure, so that part of it escapes through the soil, suppressing the influx of respiratory CO_2 from this source. Evaporation is also suppressed in this way, provided that the soil is covered by a 3-cm layer of gravel.

The rate of input of outside air is adjusted with respect to the rate of CO_2 exchange with the crop and additional CO_2 may be added at the inlet. Experiments with high CO_2 concentrations are also possible. The CO_2 concentration of the incoming air and the air in the enclosure, is measured with an infrared gas analyser. Except inside the enclosure, the air circuit does not leak, so that with thorough mixing, the CO_2 concentration within the enclosure is the CO_2 concentration of the outgoing air.

The cooler is designed in such a way that the condensed water drops with little obstruction into the vessel of an automatic raingauge. Together with the measurement of the absolute humidity of incoming and outgoing air with regularly calibrated lithium chloride cells, the transpiration rate is determined for hourly periods. Because of the time lag in collecting the water, a higher resolution time is not possible. Recently, it became possible to measure the transpiration with a resolution of the order of minutes by measuring the rate of circulation of air and water vapour differences between air entering and leaving the enclosure through the circulation system; humidities are also measured with an infrared gas analyser.

Plate 1 gives an impression of the installation which is so mobile that it can be put to work within a few hours. The power for the

Plate 1. The mobile installation of the Centre for Agrobiological Research (Wageningen) for measuring CO_2 assimilation, respiration and transpiration of crop surfaces.

installation is supplied by a 20 kW generator, which is placed at a considerable distance from the mobile laboratory and preferably downwind to avoid the CO_2 from the exhaust of the diesel engine reaching the air inlet at the top of the tube on the van.

The weather within the enclosure is, of course, different from the weather outside and it would be futile to try and make them the same. Instead, the weather section of BACROS is adapted. The influence of the transparent cover on the absorption of short-wave radiation and on heat radiation is included. It is taken into account that no wind profile is formed, the air within the whole enclosure being thoroughly mixed. The influence of the characteristic time-lags of the instrumentation on the measured CO_2-contents are also included.

3 Weather

The crop growth models simulate situations which are characterized by optimal supply of water and nutrients. Under these conditions weather is the main determinant of growth so that only weather parameters have to be introduced in the form of forcing functions. This method does not present any difficulty for the experimental evaluation of the simulation program. The appropriate weather factors, like temperature, wind speed, incoming radiation, may be measured at some arbitrary horizontal boundary in the air above the crop and the weather parameters within the crop and the soil can be simulated from these (Goudriaan, 1977). Simulated data and data actually measured within the crop and soil can then be compared to evaluate the micrometeorological aspects of the simulation.

Simulation programs are not only constructed for evaluation, but also for application. For instance, if irrigation schemes are planned in an arid region, it may be useful to predict yield under optimal conditions. However, the meteorological data used for this prediction concern the region before and not after the irrigation works have been set up. With the introduction of irrigation, the temperature at two meters is lower, the humidity higher and the wind speed is reduced; especially the net long-wave radiation will be affected by the temperature of the underlying radiation surface. Then it is impossible to introduce some boundary at which measured data may be correctly used as forcing functions, that is as data that are independent of the conditions at the soil surface. Hence in a final analysis, simulation of crop growth is only possible by simulating the weather pattern on a macro-scale.

Of course, a compromise is possible. The simplest approach to a practical solution is to assume that except for long-wave radiation the effect of the surface condition on the measurements at standard screen height is negligible and this assumption is in general made for conditions in the Netherlands where the soil surface in meteorological stations is covered with grass, reasonably supplied with water and where most fields are also under green cover. This course is also taken in arid regions, where crops are grown on a relatively small scale. However, where large-scale irrigation projects are anticipated, a meteorological analysis to estimate expected changes in weather parameters seems necessary.

Such an analysis is not attempted here, except for one aspect, which concerns radiation exchange. As long as cloudiness is not affected, it is fair to assume that the total global (short-wave) radiation is independent of the condition at the soil surface. However long-wave radiation will certainly be affected and because net radiation determines to a large extent evaporative demand, it is dangerous to use data on net radiation under conditions other than those under which they are measured. A proper course of action may be the computation of sky temperature from short-wave radiation, net radiation, and surface temperature and to use this sky temperature as a forcing function. But this computation can be done only when the station's measurements are complete and accurate.

There are, however, also many conditions where it is necessary to rely on Angström's and Brunt's type of formula for the computation of the incoming short-wave radiation and the long-wave radiation exchange. Then any sophistication is worthless, even in subsequent simulations of the micro-weather.

3.1 Input weather data

The main weather parameters show a distinct daily course and since the plant response to environmental factors is obviously non-linear and many processes are interacting, it is necessary to account for this systematic daily course of weather parameters in simulation programs that reflect the main physiological processes of plants. However it is not worth the effort to include in the weather input all details of the weather pattern that may be recorded with sophisticated instrumentation. Moreover, it is practical to work with standard meteorological weather reports.

Therefore, a program section was developed which generates daily courses of weather data from daily totals and daily maximum and minimum values of weather parameters measured at standard screen height, using the latitude of the location and the date as further information. These program sections concern incoming short-wave radiation, long-wave radiation, temperature, humidity and wind speed.

3.1.1 *Incoming short-wave radiation*

Both photosynthesis and transpiration of leaves may respond non-linearly to radiation, so that not only the absolute course of radiation intensity throughout the day, but also its distribution over direct and diffuse radiation has to be approximated. The incoming radiation flux during a time interval of simulation is then computed from the radiation fluxes with a clear and overcast sky by assuming that the sky is overcast for the time fraction (f) and clear for the time fraction $(1-f)$. Hence, it is assumed in BACROS that the clouds are evenly distributed over the day.

The basic data for the computation are presented in Fig. 2 which gives the radiation flux in the 400–700 nm wave band (visible range) in Joule $m^{-2}\,s^{-1}$ on a perfectly clear or overcast day dependent on the height of the sun. The flux on the perfectly clear day is again separated into direct radiation flux from the sun and diffuse sky radiation. The values have to be multiplied by 2 and 1.7 to obtain the short-wave radiation over the whole range of wavelengths for a clear and an overcast sky, respectively (de Wit, 1965). The radiation on an overcast day is 0.20 times the radiation on a perfectly clear day. In Angström's formula (1924), which relates radiation flux to cloud cover, this ratio is close to 0.25. However this formula refers in general to measured radiation fluxes on clear days and these fluxes may be 10–20% lower than reported here because of dust and water vapour. Perfect clear skies are used here to avoid computation of maximum radiation totals that are smaller than measured totals. The observational data are summarized in function tables for direct radiation on clear days from the sun, diffuse radiation from the sky on clear days and diffuse radiation on overcast days.

These data for standard clear and standard overcast skies, integrated to obtain daily totals for a clear and an overcast day, are $\int S_c$ and $\int S_o$, respectively. Total daily radiation actually measured $\int S$ is related to the daily totals of clear and overcast skies to compute what is called the fraction of time that the sky is overcast

$$f = \frac{\int S_c - \int S}{\int S_c - \int S_o} \qquad (3.1)$$

Now the current total short-wave radiation at any moment of the day is estimated with this fraction of the sky that is overcast according to

$$S = fS_o + (1-f)S_c \qquad (3.2)$$

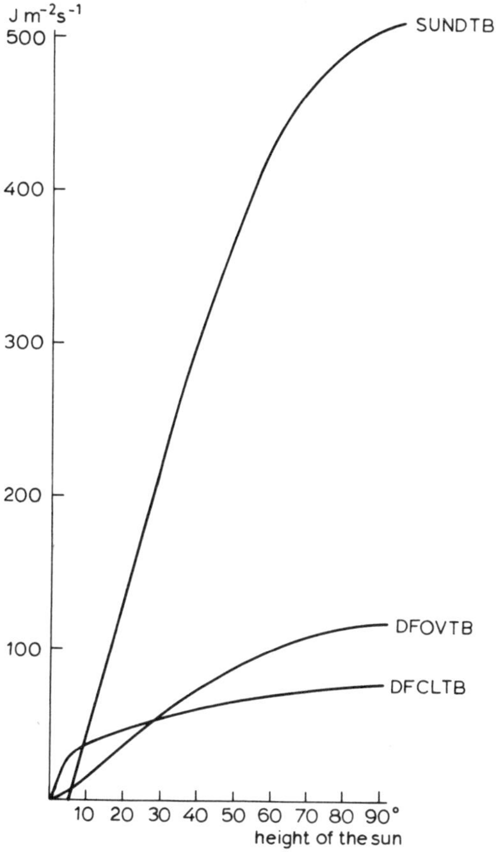

Fig. 2. Incoming visible (photosynthetically active) radiation (400–700 nm) as a function of the solar height. SUNDTB stands for the direct flux on a standard clear day, DFCLTB for the diffuse sky radiation on a standard clear day and DFOVTB for the diffuse radiation on a standard overcast day.

The standard fluxes of short-wave radiation for an overcast and a clear sky, S_o and S_c respectively, are calculated from the sine of the height of the sun. This sine depends on the sine and cosine of declination and latitude and the cosine of the hour angle of the sun according to

$$\sin\beta = \sin\lambda \sin\delta + \cos\lambda \cos\delta \cos 2\pi(t_h + 12)/24 \tag{3.3}$$

in which

β = height of the sun
λ = latitude of the site
δ = declination of the sun
t_h = hour of the day, time expressed in hours.

The declination δ is given by (expressed in radians):

$$\delta = -\frac{23.4\,\pi}{180} \cos 2\pi(t_d + 10)/365 \tag{3.4}$$

where t_d is the number of days since 1 January.

The above procedure is followed when daily radiation totals are measured. Otherwise they may be calculated with Angström's formula:

$$\int S = \left(a + b\frac{n}{N}\right) Q \tag{3.5}$$

In this formula the number of hours of sunshine (n) is often known from measurements with Campbell–Stokes sunshine recorders, but sometimes only estimates are available. N is the daylength in hours, a and b are factors depending on location and are of the order of 0.25 and 0.75, respectively. Q is total daily radiation with a clear sky. As has been said, this total radiation is often 10–20% lower than the radiation obtained from the data in Fig. 2, which hold for skies with hardly any dust or water vapour. Hence, if the data of this table are used to compute Q and then $\int S$, the value of a and b should be adjusted. A careful analysis of basic observational data and computational procedure is necessary if systematic errors larger than 10–20% are to be avoided.

3.1.2 Long-wave radiation

The simplest way to estimate loss of long-wave radiation (Joule $m^{-2}\,s^{-1}$) is by Brunt's (1932) formula

$$B_n = -\sigma T_{abs}^4 (0.56 - 0.092(0.75 e_a)^{0.5})(1 - 0.9f) \tag{3.6}$$

in which T_{abs} is the absolute air temperature, σ is the Stefan–Boltzmann constant, e_a is the vapour pressure in mm Hg, and f is the fraction of time that the sun is obscured by clouds.

Since the temperature of the crop surface is also simulated, it is tempting to substitute this temperature for the temperature of the air. However Brunt's formula is based on an analysis of experimental data, so that such a substitution cannot be done without changing some of the constants. Since air and crop temperature do not differ much for closed crops well supplied with water, this substitution is anyhow not worthwhile.

As for short-wave radiation, B_n is calculated separately for a clear and an overcast sky. According to Eqn (3.6), B_n for an overcast sky is 0.1 of the value for a clear sky. The calculation of the fraction of overcast sky f has been explained in 3.1.1.

The main problem of using Brunt's formula is not the exact value of the constants, but the assumption that cloudiness during the night is the same as that during the day, which is unlikely. Only by measuring sky temperatures, can the calculation of loss of long-wave radiation be significantly improved.

3.1.3 Temperature, dew point and wind speed

In general, maximum and minimum temperatures are available for each day and these are used for reconstructing the daily courses. For this purpose, it is assumed that the maximum occurs at 14h00 and the minimum at sunrise. The daily course is described by a sine wave for the period from sunrise to 14h00 and another sine wave for the period from 14h00 to sunrise on the next day.

The same procedure is followed for the dew point. However, the calculated dew-point may be higher than the calculated temperature because of schematization errors or errors in measurements. To avoid these, a lower limit equal to the air temperature is introduced for the dew point.

The daily wind run is measured at screen height. The wind speed is assumed to be twice as high during the day as during the night in such a way that the total daily wind run is equal to the value measured. The wind speed at the top of the canopy may be calculated from the logarithmic wind profile and the wind speed at standard height.

The computation of the turbulent diffusion resistance is based on the assumption of a logarithmic wind profile, an expression for the turbulent diffusion resistance in s m^{-1} between the crop and the air derived by

$$r_h = \ln\left(\frac{z_r - d}{z_o}\right)\ln\left(\frac{z_r - d}{z_c - d}\right) \Big/ (k^2 u_r) \tag{3.7}$$

in which u_r is the wind speed in m s^{-1} at a height z_r above the crop surface, k the Von Karman constant (0.4) and z_c is the height of the crop. The stability correction factor is assumed to be one, because it is also assumed that the air temperature within and above the crop is the same. The heights d and z_o are the zero plane displacement and the roughness length of the crop, it being assumed that the wind speed is 'zero' at a height $d + z_o$.

Considerable experimentation has been done to determine d and z_o. Since canopies are not rigidly constructed, both depend on wind speed (Monteith, 1973). If this dependence is neglected, it is often assumed that d and z_o are only proportional to the height of the crop according to $d = 0.63 z_c$ and $z_o = 0.13 z_c$. The wind speed should be measured at a height of at least one metre above the crop, and sufficient fetch should be taken into account. As has already been said (2.1), the wind speed measured at standard height in meteorological stations is taken as a substitute. For computation of wind speed within the canopy, wind speed at the top of the canopy (u_c) is needed. If one assumes the wind profile to be logarithmic,

$$u_c = u_r \ln\left(\frac{z_c - d}{z_o}\right) \Big/ \ln\left(\frac{z_r - d}{z_o}\right) \tag{3.8}$$

In crop enclosures, the air is kept in turbulent motion and the turbulence of the air is the same throughout the canopy. This difference between crops in enclosures and normally exposed crops may lead to considerable differences in transpiration rates.

3.2 Micro-weather

3.2.1 Micro-weather simulators

Temperature, humidity, wind and radiation in a crop change with time and height and to simulate their time course, distributive models are necessary in which both time and height are discretized. Such models are conveniently represented as an electrical network consisting of capacitors and resistances, as in Fig. 3. The capacitors in the centre of the figure represent the heat capacity of the leaves in each stratum. The resistances at the left represent the boundary layer resistance to the flow of sensible heat and the resistances on

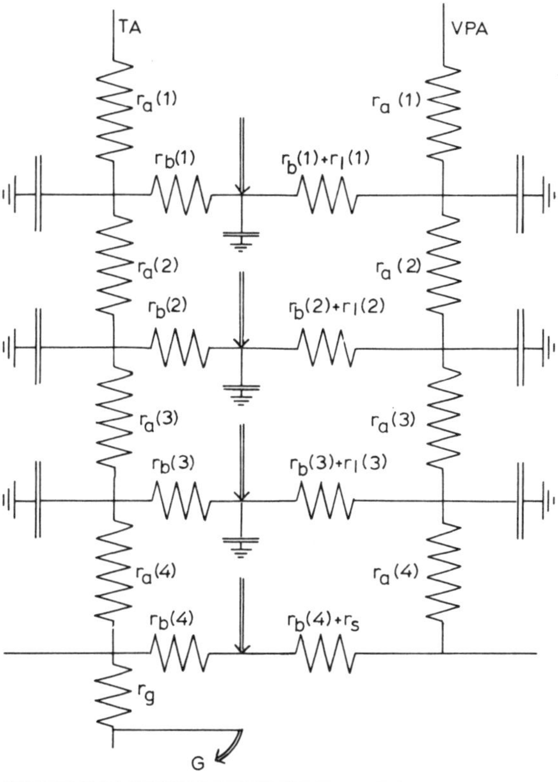

Fig. 3. A scheme of the resistances and capacitors for sensible and latent heat exchange inside a vegetation. The arrows represent the absorbed net radiation in the leaf layers and on the soil surface. See text for further details.

the right the stomatal and boundary layer resistances to the flow of latent heat in the form of water vapour. The resistances and capacitors in series on the left side of Fig. 3 stand for the exchange resistances and heat capacities of the air within the layers and on the right side for latent heat flow. The sensible and latent heat contents (the capacitors) are represented by integrals and the rates of flow are proportional to temperature or vapour pressure differences, the conductivities (inverse of the resistances) depending on leaf mass per layer, stomatal behaviour, wind speed within the layer and so on. The system is forced to function by radiation exchange within the crop and by temperature, humidity and wind speed above the crop. The transfer processes of heat, water and nutrients within the soil can be simulated in a similar way, the interface with the air being formed by a thin layer of soil at the surface. As the flow of sensible and latent heat is very large compared with the storage capacity of leaves and the air between leaves, the time constant of the system is of the order of only 100 seconds. Hence, it is impossible to operate the model for a season or even a day. This problem can be solved by application of Goudriaan's (1977) bypassing method for stiff systems.

It is also possible to eliminate the capacitors so that only a network of resistances remains. A solution may then be obtained by matrix algebra. However such programs are far less lucid, more difficult to change, and simplifications and linear approximations have to be introduced to obtain a straightforward solution, (Goudriaan, 1977).

The macro-weather data that are needed for the operation of such a detailed model are the temperature and humidity of the air, the wind speed, the direct and diffuse radiation from the sun and the sky temperature or net radiation. Values measured above the simulated canopy should be used in detailed models. It is a matter of further analysis to what extent the use of data of normal weather stations is justified.

The foliage characteristics that must be known are the leaf area index (LAI), the leaf width and the canopy architecture, the extinction of visible, short-wave and net radiation and the extinction of wind and exchange coefficients. The plant physiological characteristics exert their influence mainly through stomatal conductivity which governs the division of incoming radiant energy into sensible heat and latent heat of evaporation. Stomatal conductance in its turn does not depend only on the water status of the plant, but also on CO_2 assimilation. The main soil characteristics are the hydraulic and thermal properties, which may be estimated from the composition and type of soil or they can be directly measured. Moreover some average clod dimension is necessary to characterize the roughness of the soil surface. The wetness of the surface again determines the division of radiation into sensible and latent heat and a reasonable simulation of the soil moisture content of the surface is necessary for operating the model.

Goudriaan (1977) verified the operation of his simulator by comparison with a series of measurements of Stigter *et al.* (1977).

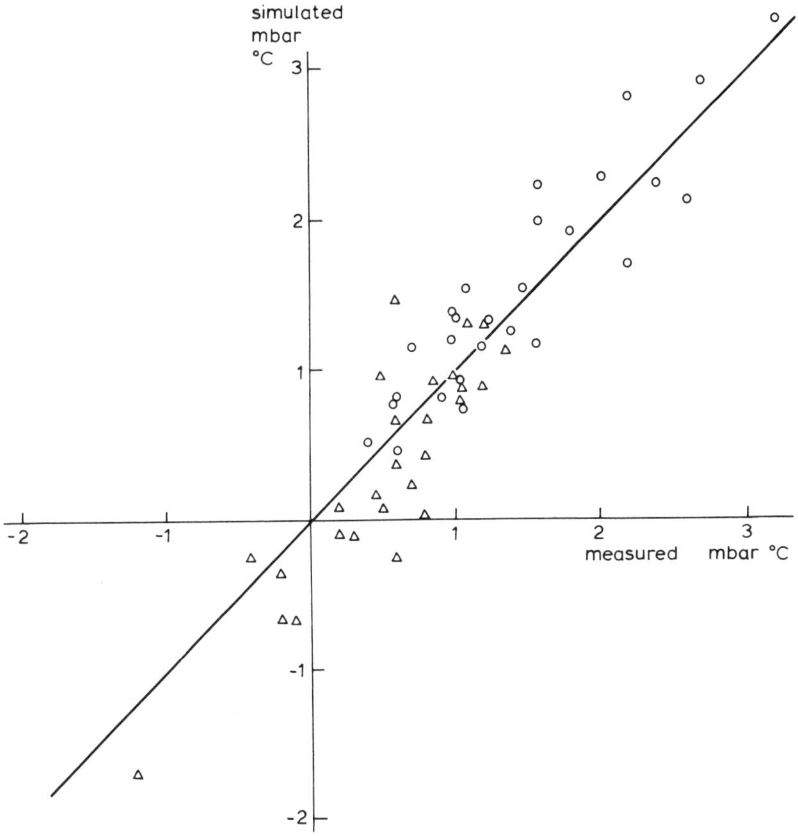

Fig. 4. Simulated and measured differences between the temperatures (Δ) and humidities (○) inside a canopy of maize and those above.

The performance is reasonable, as is shown in Figs 4 and 5 for a maize crop, reasonably supplied with water. Fig. 4 concerns the measured and simulated differences between the humidity inside and above the crop, whereas Fig. 5 concerns temperature and humidity profiles as measured and simulated on two days. Neither the measured nor the simulated temperatures and humidities in the crop differ by more than 3°C or 3 millibars from those above the crop.

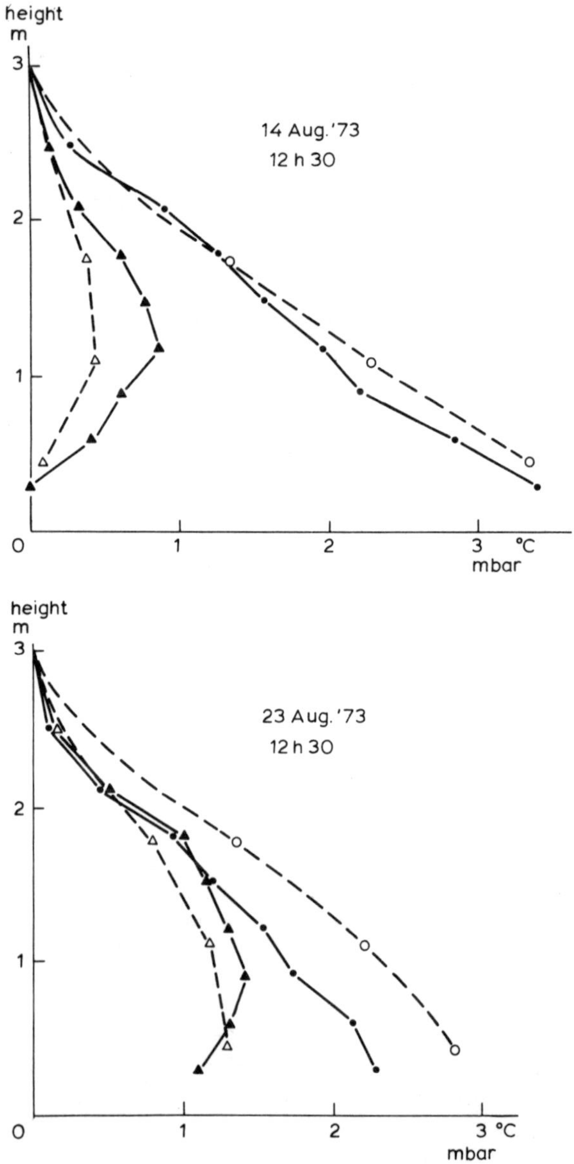

Fig. 5. Simulated (broken line) and measured (solid line) profiles of air temperature (Δ) and humidity (O) inside a maize crop.

Since the responses of physiological processes to temperature are often not known with great precision, it seems hardly worthwhile to make all this effort for the simulation of small temperature differences. On the other hand, the period of dew formation at different heights in the crop may vary considerably, so that in studies of the epidemiology of fungus diseases, for instance, detailed models are justified. Such models are also justified for open canopies and under conditions where the water status of the soil is suboptimal and there is such a feedback on the water status of the plant that the stomatal behaviour is affected.

3.2.2 A simplified micro-weather simulator

The macro-weather data, the knowledge of the plant physiological responses, and the purpose of BACROS do not justify the use of detailed micro-weather simulators.

Considerable simplification is achieved by assuming that the turbulent resistance in the vegetation is zero, so that the temperature, humidity and CO_2 concentration of the air is the same throughout the canopy. This simplification is suggested by the small gradients that exist in conditions where the crop is well supplied with water and the soil surface is not dried out (Fig. 5). It eliminates the sensible and latent heat capacity of the air and the interactions between the leaf strata within the crop. Another simplification is the use of Penman's (1948) combination method to compute the transfer of sensible and latent heat from the leaf strata to the surroundings, so that the heat capacity of leaves, another series of integrals with small time constants, is eliminated.

In Fig. 3 this simplification is introduced by neglecting the turbulent resistances inside the canopy ($r_a(2)$, $r_a(3)$ and $r_a(4)$), so that the conditions of the air inside the canopy are uniform. In Table 1 the effect of this simplification is presented. The differences between the simulations are very small, so that for simulations with the crop growth simulator the use of only one air layer is fully acceptable. Within this simplified model three kinds of resistances are needed: the turbulent diffusion resistance of the air (r_a), the boundary layer resistance of the leaves (r_b) and the stomatal resistance of the leaves (r_l). The first two resistances, expressed in the correct units, are almost the same for sensible and latent heat (water vapour). The resistances r_b and r_l vary with the depth of the leaves in the canopy. The boundary layer resistance depends mainly on the wind velocity around the

Table 1. Simulated net CO_2-assimilation NCASC, total latent heat loss TEHL, total sensible heat loss TSHL and soil heat flux G for a maize crop at 12h00. The first column gives results when the profiles inside the vegetation are taken into account. In the second column they are neglected (Goudriaan, 1977)

fluxes at noon	dimensions	with profile inside vegetation	turbulent resistance inside vegetation zero
NCASC	kg CO_2 ha^{-1} h^{-1}	83.8	91.0
TEHL	J m^{-2} s^{-1}	320.2	322.5
TSHL	J m^{-2} s^{-1}	263.2	258.5
G	J m^{-2} s^{-1}	70.4	70.2

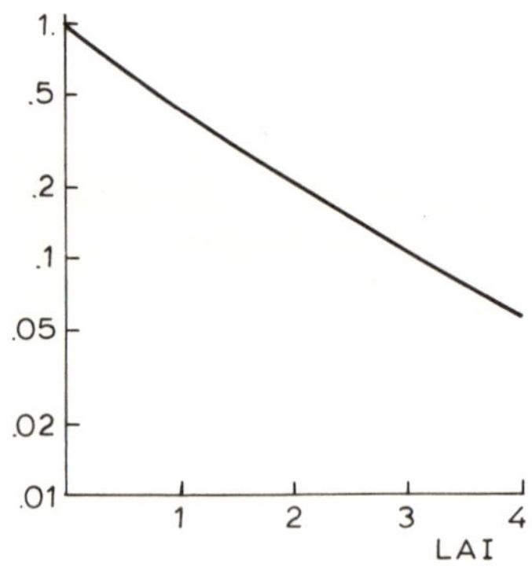

Fig. 6. Calculated extinction of diffuse radiation for a crop with black leaves and a spherical leaf angle distribution. The extinction is almost exponential.

leaves, so that for its computation the extinction of wind speed in the canopy is needed. The resistance of the stomata governs the ratio between sensible and latent heat loss of the leaves and will therefore be treated in considerable detail in 5.2. The 'wetness' of the soil surface is described by the resistance for evaporation r_s, which has the same controlling function for water loss as the leaf resistance r_l. The soil heat flux G passes through the resistance r_g from the soil surface to the centre of the top soil layer.

Light intensity decreases exponentially with depth in the canopy with a different extinction coefficient for visible, near-infrared and thermal radiation. It is possible to arrive at simple expressions for transmission and reflection coefficients of canopies with arbitrary leaf distributions (Goudriaan, 1977). The radiation from the sun but with a black sky (as on the moon) is then considered first for a canopy with a large leaf area index. The extinction for black leaves with a spherical leaf distribution is given by a solid line in Fig. 6. Since no radiation is reflected, this line also represents the net downward flux. With scattering leaves, some light is reflected so that the net flux at the top of the canopy is less. On the other hand the net flux reaches further downwards because of the scattering.

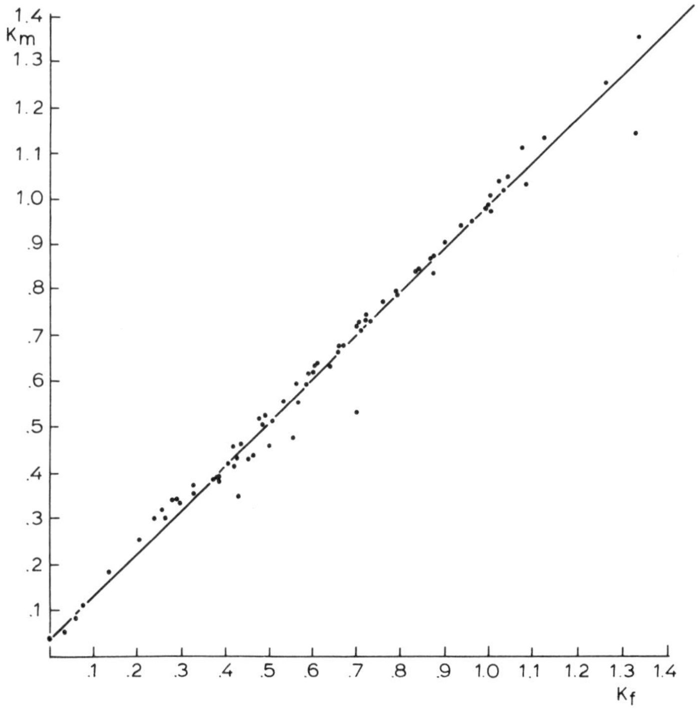

Fig. 7. Numerically computed extinction coefficients (K_m) versus extinction coefficients (K_f) calculated with a simple equation (3.4), for many situations differing in leaf angle distribution, scattering coefficient and geometry of the incoming radiation.

Subsequent analysis of numerical data obtained for a wide range of leaf distribution functions, scattering coefficients and geometry of the incoming radiation showed that the resulting extinction coefficient in the function $e^{-K(\beta)\mathrm{LAI}}$ may be fairly well approximated by the expression

$$K(\beta) = (1-\sigma)^{0.5} K_b(\beta) \tag{3.9}$$

in which σ is the scattering coefficient of the leaves and $K_b(\beta)$ the extinction coefficient of black leaves computed for radiation from direction β.

Good agreement is shown in Fig. 7 where numerically computed values of K are plotted against the values calculated with Eqn (3.9) for more than 200 cases. With the extinction coefficient the net flux of radiation at a certain depth can be calculated. This flux consists of a downward flux of direct radiation and downward and upward fluxes of diffuse radiation. Subtraction of direct flux from the total, gives the diffuse part of net flux.

These formulas account only for the radiation from the sun. The radiation from the sky has to be superimposed on this. The extinction coefficient for diffuse radiation can be found by the weighted summation for the nine zones of the upper hemisphere of 10 degrees each.

Expressions of similar complexity are developed also for reflection coefficients of crop surfaces, but for these the reader is referred to the original publication of Goudriaan (1977) and to the listing of the simulation program.

4 Water status of crop

The water status of the canopy is governed by the balance between water loss through transpiration and water supply by the roots. Both are adjusted to each other through the relative water content of the canopy, which may affect on the one hand the opening of the stomata and on the other the difference in water potential that governs water uptake. Several attempts have been made to treat the relation between the weight and the geometry of the root and the conductance for water uptake (Brouwer & de Wit, 1968; Lambert & Penning de Vries, 1973). However these aspects are not incorporated in detail in the present model because emphasis here is on the canopy and its growth.

4.1 Transpiration

The rate of transpiration of the crop is obtained by summing the rates of transpiration of successive leaf layers with a given leaf area index. The calculations for each leaf layer are based on the combination method first proposed by Penman (1948). In this approximate solution, storage of heat in the transpiring leaf is neglected. The expression for the latent heat loss per unit leaf area is then

$$\lambda E = (sR_n + (e_s - e_a)\rho c_p / r_b)/(s + \gamma(r_b + r_l)/r_b) \qquad (4.1)$$

in which e_s and e_a are the saturated and actual vapour pressure, respectively, s is the slope of the curve of saturated vapour pressure against temperature at air temperature, γ is the psychrometer constant, R_n is the absorbed radiation, r_l and r_b are the resistances of the leaf and its laminar layer, and ρc_p is the volumetric heat capacity of the air. This equation is used in the simulation model to calculate the transpiration from individual leaf layers. The radiation absorbed in each leaf layer follows from the balance between incoming and outgoing radiation (3.1). For each of the radiation components, exponential extinction with depth in the canopy is assumed, each one with its specific extinction coefficient. These coefficients are calculated from the crop architecture and the angle of the sun. The energy consumed by photosynthetic activity is also taken into account.

The resistance to water vapour exchange through the laminar layer, r_b, is dependent on the wind speed around the leaves and the size of the leaves. Several semi-empirical relations have been proposed to calculate the resistance from these variables. The preference for any particular formula depends on the conditions under which it was established and those under which it should be used. For the present purpose the most reasonable approximation seems to be $r_b = a\sqrt{w/u}$, in which a is a constant, w the width of leaf and u the wind speed (described by Pearman *et al.*, 1972). The wind speed in each leaf layer is obtained from that above the canopy, assuming exponential extinction, with an extinction coefficient of 0.7; for this wind speed a lower limit of 0.02 m s^{-1} is assumed. Stomatal conductance is governed by the CO_2 concentration in the stomatal cavity, within given boundaries and is thus calculated from the current rate of assimilation (5.3). It is, however, assumed that under conditions of water stress, stomatal control by CO_2 is ineffective. Then conductance is governed by the water status of the canopy, increasing dehydration being accompanied by gradual closure of the stomata. Since cuticular conductance is not negligible with nearly closed stomata, it is added to the stomatal conductance to obtain total conductance of the leaf surface. When the evaporative heat loss is known, sensible heat loss is calculated by

$$C = R_n - \lambda E - M \qquad (4.2)$$

in which M is the energy consumed for photosynthetic activity. The temperature of the leaves T_l follows now from

$$T_l = T_a + Cr_b/(\rho c_p) \qquad (4.3)$$

in which T_a is the temperature of the air.

The calculations for each leaf layer are done separately for the fraction of time that the sky is clear $(1-f)$ and for the fraction of time that the sky is overcast (f). During the time that the sky is clear, a distinction is made between leaves that are exposed to direct sunlight, the sunlit leaf area, and leaves that receive only diffuse radiation, the shaded leaf area. Total transpiration of the canopy is obtained by summing the energy fluxes for all leaf classes, as is the total sensible heat loss. The average temperature of the crop follows from the average temperature of the various leaves.

4.2 Water uptake

Uptake of water from the soil is governed by the difference in water potential in the crop and in the soil and by the resistance to water flow in the soil-plant system. Detailed models are available, in which the flow of water through the soil and the expansion of the root system are taken into account (Lambert & Penning de Vries, 1973). However such a detailed treatment is unwarranted in the present program because optimum soil moisture conditions are assumed throughout and because our knowledge about growth and functioning of the root system is fragmentary. Moreover sufficiently accurate methods of experimentation have only recently been developed. In the present model therefore, the root system is treated as one unit, with respect to water flow characterized by its conductance.

The water potential in the soil is maintained at −0.1 bar, which corresponds approximately to field capacity. The water status of the canopy is characterized by its relative water content and its total water potential. For the time being a unique relation between the two is assumed. This relation may vary as a result of variations in the component potentials contributing to the total potential i.e. turgor pressure and osmotic potential. Especially the latter may show variations due to chemical transformations in the plant. Lack of quantitative data, however, restricts the application of these aspects in the simulation model. A direct relation between relative water content of the crop and water potential of the crop is assumed, neglecting temporary changes of osmotic pressure. Measurements of Kleinendorst & Brouwer (1972) show that this assumption may be fair.

The total resistance to liquid flow through the plant is assumed to be primarily concentrated in the root system, where the water must traverse the protoplast as cell walls are suberized. There is also a resistance to water flow in the xylem vessels. Such a resistance would presumably be dependent on the size of the conducting tissue and thus on the weight of the crop. In the present approach this resistance has been neglected, also because it is generally accepted that its influence is of little practical importance.

4.3 Root resistance

The literature on the nature and magnitude of root resistance and the factors influencing it is voluminous and often conflicting. Generally a distinction is made between active and passive transport of water across the outer cell-layers of the root to its xylem vessels. However most authors agree that the potential gradients developing under influence of transpiring leaves are such, that only passive transport along these gradients is of practical importance. Research has therefore been largely restricted to phenomena associated with this transport and then mainly on individual plants in nutrient solutions. The applicability of the results in simulation models referring to the field situation is thus limited. In the model the conductance of the root system is determined from the weight of the root system, its composition i.e. the ratio between 'young' and 'old' roots, the soil temperature and a conversion factor relating weight to conductance.

The growing roots are accumulated in an integral with young roots, while at the same time these roots are subject to suberization. Suberization proceeds in proportion to the amount of young roots present, with a time constant of about 5 days. The conductance of the old roots, through which water transport is hampered as a result of the suberine layers, is set at 0.3 times that of young roots. This value is an average of ratios of root conductances determined along a growing root, with increasing distance from the root tip (Brouwer, 1965). The suberized roots may die off during the growing season; the total root weight of a crop often decreases as it matures. Subsequently these dead roots may be subject to decay as a result of microbial action. Since the quantitative aspects of these processes are elusive, they have not been included in the model.

The temperature of the medium has a distinct influence on the uptake of water by the root system. For proper understanding of this phenomenon, two processes must be distinguished. In the first place temperature may influence the structure of cell membranes, thus changing the conductance of the roots. On the other hand increasing temperatures give rise to decreasing viscosity of water, which facilitates transport across the root. Kuiper (1964) demonstrated with beans that in the lower temperature range (up to ±15°C) both effects influence water uptake, resulting in a Q_{10} value of about 4, while above 15°C changes in uptake rate can be fully accounted for by changes in viscosity. The relation between temperature and water uptake introduced in the program to describe the effect of temperature on root conductance (Fig. 8) is based on own measurements with maize.

The conversion factor, relating root conductance to root weight and of primary importance for the calculation of conductance, is

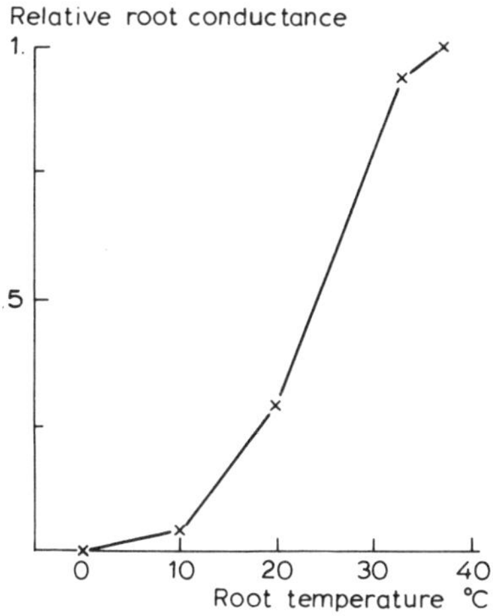

Fig. 8. Root conductance as a function of root temperature, relative to the conductance at 37°C.

difficult to obtain from experimental data. As mentioned earlier, most available data refer to single plants in pots grown on nutrient solution, in which the quantitative relations may be quite different from the field situation, even though the processes are the same. The numerical value used in the model has therefore been estimated, assuming that a crop, well supplied with water can maintain its turgidity under a fairly high evaporative demand. An LAI of 1, 50 kg young roots and 450 kg old roots are assumed to exist. At a high transpiration rate of 0.165 g m^{-2} s^{-1} and a relative water content of 0.975, the required conductance is then 0.08 g water bar^{-1} s^{-1} m^{-2}. This leads to a value for the ratio weight: conductance of ±2500 kg ha^{-1} per g water bar^{-1} s^{-1} m^{-2} which is used in the model. Surprizingly enough, this calculated ratio is of the same order as comparable data from nutrient solution experiments (Brouwer, 1965).

Another aspect of the root resistance, which has received attention in the literature is its dependence on the potential difference across the root, or the flow rate. Barrs (1970) reported that for a number of crops root resistance was inversely proportional to rate of flow so that the plant can maintain full turgidity under increasing evaporative demands (Fig. 9). However, this phenomenon was

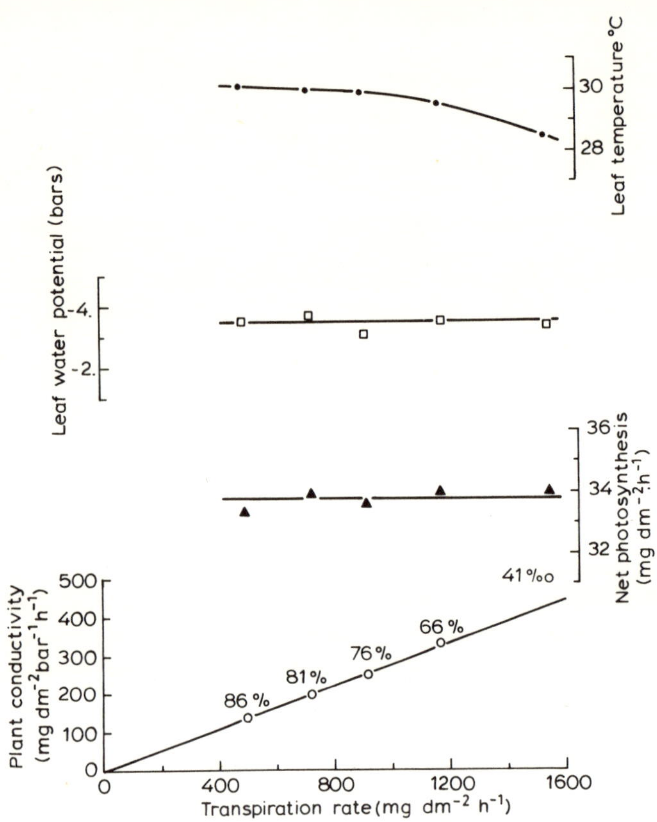

Fig. 9. Relations between maize leaf temperature, leaf water potential, conductance and transpiration rate. The last was varied by varying the relative humidity from 41–80% (Barrs, 1970).

observed under low evaporative demand, while at high evaporative demand the root resistance seemed constant. Kuiper (1972) explained this behaviour by assuming that new pores for moisture transport are formed in the plasma membrane as a result of increasing pressure. He also reported that it is restricted to pressures of 1 to 2 bar after which a linear relation is observed between pressure and water uptake. In the present model this phenomenon has therefore been neglected.

4.4 Water balance

The term water balance as used here refers to the balance between water lost by transpiration and that taken up by the root system. The description given so far is identical for the two versions of the simulation model, the first one referring to a whole growing season, and the second one calculating the daily course of photosynthesis, respiration and transpiration. The main difference between the two versions of essentially the same model, is the time interval with which the programs are executed. The second version is executed with variable time intervals, the magnitude of these being determined at each moment by the 'fastest' process i.e. that with the smallest time constant (de Wit & Goudriaan, 1974). To maintain stability, the time intervals of integration are of the order of minutes or even seconds. For simulating a 24-hour period such small intervals are feasible but application for a whole growing season, involving 150 or more days, would lead to prohibitively high computing costs. The version applied to the latter situation is executed therefore with fixed intervals of one hour.

As the state variable with the smallest time constant, i.e. the heat content of the canopy, has already been eliminated by calculating canopy temperature from the radiation balance (4.1), the water content of the crop is the determining state variable. When the same example from the previous section is used, with a leaf weight of 1000 kg and a dry matter content of the material of 15%, the total amount of water is 850 kg ha^{-1}. Applying the same rate, 1.65 kg ha^{-1} s^{-1}, the time constant equals 500 s (850/1.65). Thus it can easily be seen that the model would be unstable when time intervals of one hour are applied.

The same principle as applied for the temperature is used in this situation: the water content of the canopy is not considered as a state variable. Instead it is assumed that at each moment equilibrium exists between the amount of water lost through transpiration and that taken up from the soil. The actual value of the equilibrium point is determined by the water potential or the relative water content in the canopy. The stomatal resistance is either governed by the control of the internal CO_2-concentration (5.2) or the relative water content. In the latter case, a decrease in the relative water content may cause an increase in stomatal resistance and hence a decrease in transpiration. At the same time the rate of water uptake is increased. In the model the equilibrium situation of water uptake and transpiration is found by an iterative procedure. First uptake

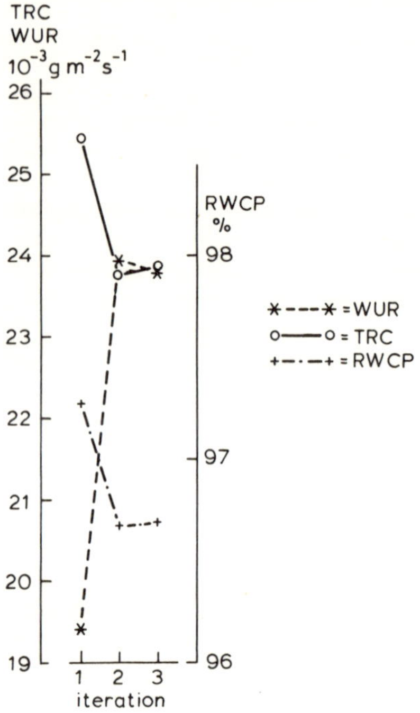

Fig. 10. Computed values of the transpiration rate TRC, the water uptake WUR, and the relative water content of the crop RWCP in subsequent iterations. When TRC and WUR are sufficiently equal, iteration is terminated.

and transpiration are calculated under the assumption that the relative water content of the previous time interval is still the equilibrium value. When evaporative demand or root conductance have changed to such an extent that the difference between uptake and transpiration is larger than a preset accuracy criterion, a new value for the relative water content is obtained dependent on the sign of the difference and the calculation of uptake and transpiration is repeated with new values for the water potential and the stomatal resistance. This procedure is repeated until the accuracy criterion is met, usually within 3 to 4 iterations. The procedure is graphically illustrated in Fig. 10. When the equilibrium value of the relative water content is obtained, the remainder of the program is executed.

5 Autotrophic processes and stomatal control

The process of CO_2 assimilation is the most important photosynthetic process. Its dependence on light is described in the program on the basis of a response curve which is characterized by the maximum rate of assimilation at high light intensity, the initial efficiency of assimilation and the dark respiration. Our group (Sinclair *et al.*, 1977) is trying to estimate these parameters from basic characteristics of the photosynthetic machinery, leaf structure and so on, but it is felt that the results are not sufficiently accurate as yet to be incorporated in growth models. The values of the parameters are therefore derived from direct measurements of the assimilation function. The advantage of this approach is simplicity and adaptability because forcing functions are used. Its disadvantage is that any feedback of past history of leaves on assimilation can only be incorporated in an elementary way.

Superimposed on assimilation is stomatal control, which appears to work in two ways. On one hand, assimilation may be controlled by stomatal closure, mediated by water shortage and on the other hand assimilation itself may control stomatal opening. The resulting interaction between transpiration and assimilation may be satisfactorily treated by considering the CO_2 concentration in the intercellular space of the leaf.

5.1 Assimilation of carbon dioxide

A characteristic light response curve of CO_2 assimilation is presented in Fig. 11. This curve is most conveniently described by:

$$F_n = (F_m - F_d)\{1 - \exp(-\varepsilon R_v / F_m)\} + F_d \qquad (5.1)$$

in which

F_n is net assimilation in kg CO_2 m^{-2} (leaf) s^{-1}
R_v is absorbed radiant flux in the 400–700 nm range in J m^{-2} s^{-1}
F_m is maximum rate of net assimilation at high light intensities in kg CO_2 m^{-2}(leaf) s^{-1}

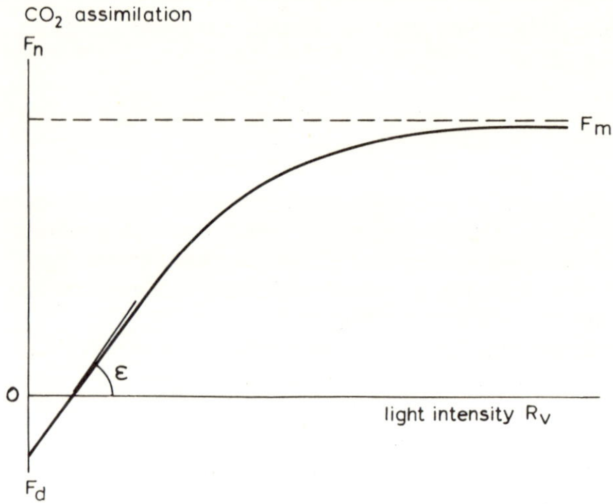

Fig. 11. A typical light response curve of the net assimilation of carbon dioxide for an individual leaf. F_d stands for the dark respiration, ε for the slope (or efficiency) at low light and F_m for the net assimilation rate at light saturation.

ε is the efficiency in kg CO_2 Joule^{-1} at the light compensation point

F_d is net assimilation in the dark in kg CO_2 m^{-2}(leaf) s^{-1} (dark respiration).

For C_3 plants, the experimentally determined number of light quanta that is necessary for the reduction of one molecule CO_2 is 15 (Björkman, 1966), but when the oxygen concentration is lower than 0.05%, only 10.5 light quanta per molecule CO_2 are required. Theoretical considerations show that the minimum number of light quanta required for reduction of one molecule CO_2 is 8 (Björkman, 1966). As the first product of assimilation is glucose with an energy content of 15700 Joule per gram, the maximum efficiency of photosynthesis is about 25% when radiation of 550 nm is used. The actual efficiency is about $\frac{8}{15} \times 25 = 13\%$, which may be explained by light absorption by other pigments than chlorophyll. In the model, the efficiency of light use in CO_2 assimilation is expressed in kg CO_2 Joule^{-1} h^{-1} ha^{-1} m^2 s. Expressed in these units an efficiency of one molecule of CO_2 per 15 light quanta has a value of:

$\frac{1}{15}\ abtm\lambda/(hcN_a)$

in which a, b and t are the conversion factors 10^{-3} kg g^{-1}, 10^4 m^2 ha^{-1} and 3600 s h^{-1}, m is the molecular weight of CO_2, 44 g, c is the velocity of light $3 \cdot 10^8$ m s^{-1}, h is Planck's constant, $6.626 \cdot 10^{-34}$ J s^{-1}, N_a is Avogadro's number, $6.0225 \cdot 10^{23}$, and λ is the wavelength taken as $550 \cdot 10^{-9}$ m. The result is 0.482 kg CO_2 Joule^{-1} h^{-1} ha^{-1} m^2 s.

Björkman & Ehleringer (1975) and Ehleringer & Björkman (1976) found that the decrease in quantum yield of C_4 plants due to their inherent higher energy requirement offsets the decrease in quantum yield of C_3 plants due to oxygen inhibition. However, this result was only for one temperature, since only the latter decrease is temperature dependent.

Our own observations are considerably less conclusive than those of Björkman *et al.*, probably because of their routine character, but possibly also due to variations in light absorption by other components than chlorophyll, in dark respiration and in other photosynthetic processes than CO_2 assimilation. Our measurements certainly do not justify the introduction of any difference in initial efficiency between C_3 and C_4 plants, or any influence of temperature within the normal range. It is assumed therefore that the initial efficiency equals 0.5 kg CO_2 Joule^{-1} ha^{-1} h^{-1} m^2 s at 300 vppm CO_2, the dependency on the latter being considered later.

The maximum assimilation F_m depends on temperature much more than does the initial efficiency, a characteristic situation for C_3 and C_4 plants being given in Fig. 12. However, this phenomenon appears to be complicated because this temperature effect depends on the pretreatment of the plants. For instance, for maize plants (not individual leaves) at a light intensity of 280 Joule m^{-2} s^{-1} and pretreated at 15°C, the CO_2 assimilation was 9.15 and 16.65 kg CO_2 ha^{-1} h^{-1} measured at 15 and 25°C, but pretreated at 25°C these values were 12.0 and 26.5 kg CO_2 ha^{-1} h^{-1} (de Wit *et al.*, 1970).

In an attempt to account for this adaptation in an elementary way, a system illustrated in Fig. 13 was developed. For maize plants grown at 25°C, the light saturated net assimilation, F_m, may depend on the temperature of the measurement as given by the curve in the upper graph. It must be noted that the maximum is reached at 32°C, the optimum temperature being higher than the temperature at which the plants are grown. When plants are grown at 20°C, and presumably adapted to this temperature, it is assumed that F_m at

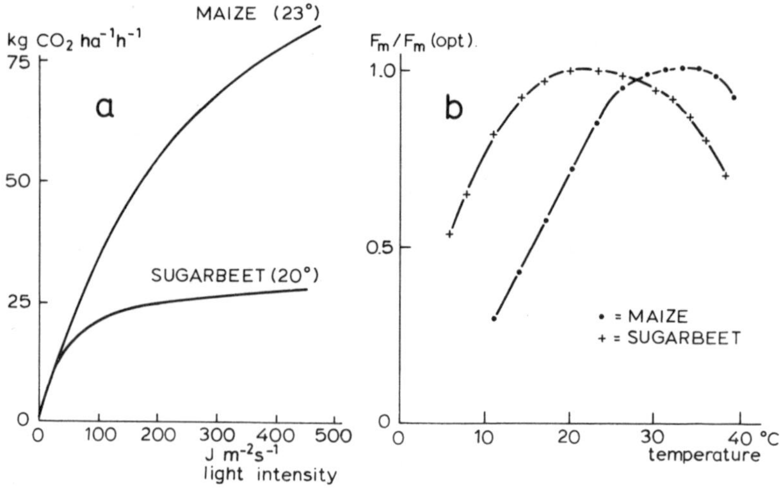

Fig. 12. (a) Light response curves of the gross assimilation of carbon dioxide for maize (C_4) and sugarbeet (C_3) leaves. (b) Dependence of the maximum rate F_m on temperature, relative to the value at the optimum temperature.

20°C for these plants is the same as F_m at 25°C for the standard plants, grown at 25°C. Further it is assumed that the minimum temperature of 10°C, where temperature response begins, is not affected. So the standard curve of 25°C can still be used, provided the actual temperature is converted to an effective temperature according to one of the straight lines given in the lower figure. The straight lines all start at the 10°C point because the minimum temperature is assumed not to change. The slope of the line is determined by the condition that the effective temperature is 25°C for the average temperature at which the plants are grown. The average growing temperature is calculated from the actual temperature in the daytime by an exponential delay with a time constant of 4 days. It seems reasonable to impose an upper and a lower limit on the adaptation capability of the plants. The upper limit for adaptation to the average growing temperature was set at 30°C and the lower limit at 18°C. The optimum temperature will be 39°C and 22°C, respectively. The standard temperature curve for the maximum assimilation of maize, as presented in Fig. 13 is derived from a representative series of measurements given in Fig. 14 (van Laar

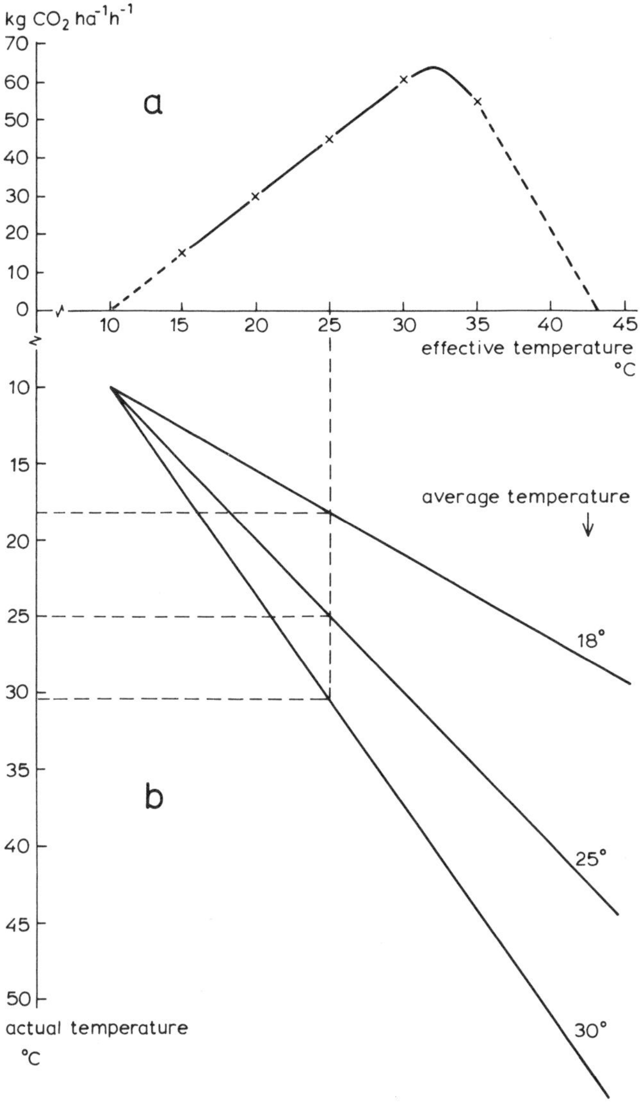

Fig. 13. (a) Temperature dependence of maximum net assimilation for maize plants grown at 25°C. (b) A method of converting actual temperature to an effective temperature by projection of the ordinate onto a straight line. The slope of this line reflects the degree of adaptation.

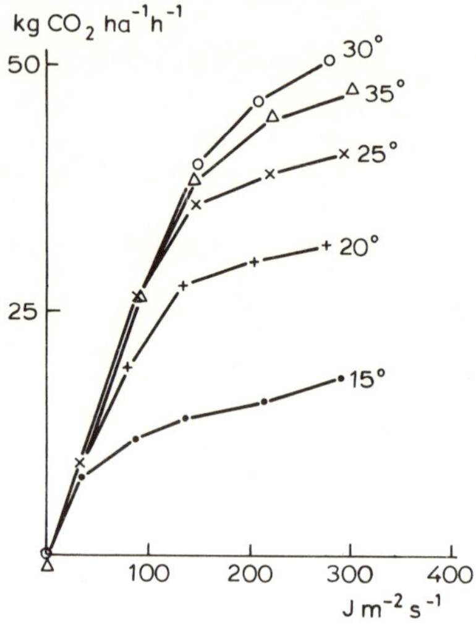

Fig. 14. Light response curves of CO_2 assimilation of maize leaves, measured at different temperatures.

& Penning de Vries, 1972) by setting the initial efficiency at 0.50 and adapting F_m according to Eqn (5.1). The part shown by a broken line was found by extrapolation.

Although this mimicking procedure may be satisfactory for summarizing the results of some experiments under laboratory conditions, it was found that under field conditions plants adapt more quickly and fully to varying temperatures. Therefore the use of this system was omitted. This evidence is so much related to field experiments with enclosures that further discussion of adaptation is postponed to Chapter 8.

5.2 Other photosynthesis processes and dark respiration

Adenosine triphosphate (ATP) is produced in the chloroplasts in the light. Certainly at higher light intensities, CO_2 may be in short supply and excess ATP may then be used for processes other than CO_2 assimilation in the leaves, such as transport of photosynthesis products, maintenance, NO_3 reduction and synthesis of amino acids and proteins. Especially the reduction of nitrate is important because it requires a considerable amount of energy and because there is much evidence at present that most of the nitrate reduction takes place in the leaves and in the light (Challa, 1976).

At lower light intensities, the magnitude of the net assimilation of carbon dioxide is to a large extent determined by dark respiration (Eqn (5.1)). The dark respiration of comparable material may vary from practically zero to over 10% of the maximum net assimilation rate in different experiments. This variation may be partly related to differences in maintenance requirements, associated with past photosynthetic activity of the leaf tissue and differences in transport rates of assimilatory products out of the leaf. Unfortunately, no physiologically justified method of determining dark respiration and photosynthesis activities other than by CO_2 assimilation is available. To proceed anyhow, the following two ad hoc assumptions were made.

Dark respiration is set equal to one-ninth of the CO_2 assimilation rate above zero, averaged with a time constant of half a day. This procedure ensures that after a period of high light intensity, this respiration is about one-ninth of the maximum net assimilation rate, as suggested by Tooming (1967) and confirmed for cotton by Mutsaers (pers. com.). On the other hand, the dark respiration of leaves exposed to low levels of light is accordingly lower, reflecting the reduced metabolic activity of these leaves. It should be mentioned here that further sensitivity analysis showed that it is not necessary for this fraction to be extremely accurate because the respiration of photosynthesizing leaves is mostly small compared with the respiration of heterotrophic tissue. In the model, dark respiration per unit leaf surface is simply set equal to one-ninth of the first order average of crop assimilation divided by the leaf area index.

The costs of maintenance, nitrate reduction, amino-acid synthesis and translocation of photosynthesis products out of the photosynthesizing leaves are not considered, as it is assumed that these processes take place during the measurements of the net assimilation of healthy leaves attached to the plants. Such a phenomenon could explain why the measured initial efficiency of light use of the leaves is in general somewhat lower than the efficiency calculated from quantum yield measurements. However, suitable methods of measuring these photosynthesis processes are not available, so that many details will remain obscure for some time to come.

5.3 Stomatal control

The above treatment should be satisfactory for a first approach if stomatal control does not interfere and the external CO_2-concentration is about 330 vppm. However even when water supply is optimal, stomatal opening may be affected by the water status of the plant and in addition CO_2 assimilation itself may affect stomatal opening. Stimulated by suggestions of Raschke (1975) these effects have been studied in considerable detail by Goudriaan & van Laar (1978), so that it suffices to treat here only the basic elements of their approach.

The CO_2 concentration in the intercellular spaces may be calculated by measuring simultaneously CO_2 assimilation, transpiration and leaf temperature. Then plants sometimes regulate the stomatal aperture in such a way that this internal CO_2-concentration is kept within narrow limits, whereas in other cases plants do not exhibit this phenomenon. An example of both situations is given in Figs 15a and b where CO_2 assimilation is varied by varying the light intensity and the external CO_2-concentration. In both figures, the net assimilation is given along the horizontal axis and the light intensity in a downward direction along the vertical axis. Hence, the observations in the fourth quadrant of the figures present the relation between net assimilation and light at different concentrations of CO_2. The inverse of the resistance to CO_2 diffusion from the external source towards the intercellular space or the conductance is given along the vertical axis in an upward direction. These resistances may be calculated from transpiration and leaf temperature data.

Fig. 15a refers to sunflower, grown under controlled conditions. Here the conductance for CO_2 diffusion is independent of the assimilation, light intensity and external CO_2-concentration. Obviously, the internal CO_2-concentration, calculated as the external concentration minus net assimilation over conductance is not regulated at all. Challa (1976) found that this regulation is also absent with cucumber and experiments with CO_2 supply in greenhouses strongly suggest that this may be a common phenomenon in horticulture.

Fig. 15b refers to maize, also grown under controlled conditions. Here conductance is proportional to net assimilation at a given external CO_2-concentration and the proportionality factor itself, that is the slope of the lines, is proportional to the external CO_2-concentration over a wide range. Thus a constant intercellular CO_2-concentration is maintained over a wide range of conditions.

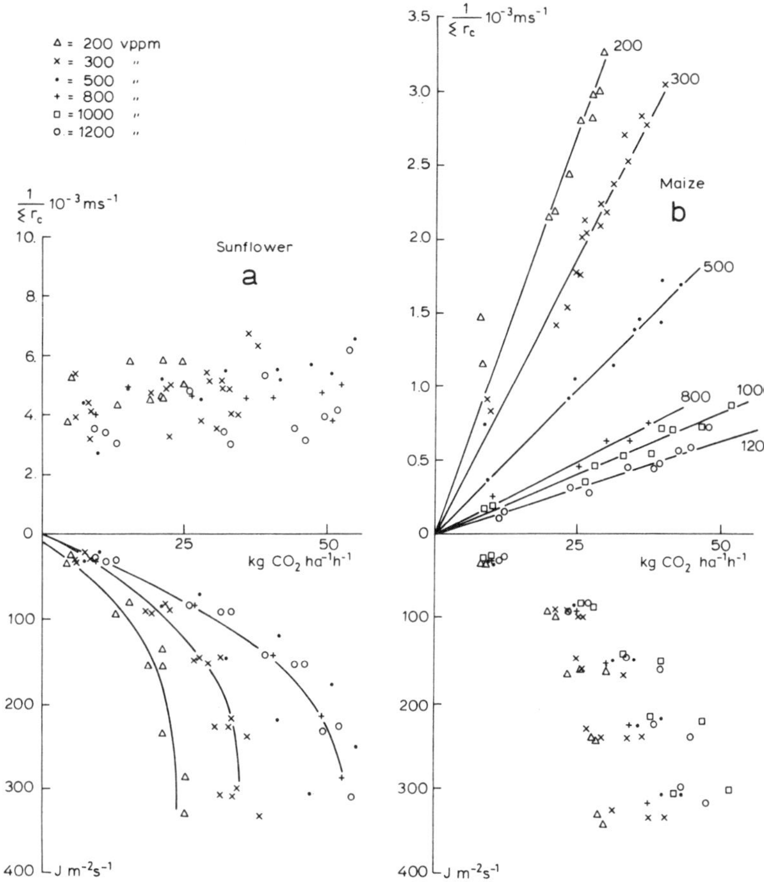

Fig. 15. Light response curves of CO_2 assimilation (lower graphs) of sunflower (a) and maize (b), measured for different concentrations of carbon dioxide. In the upper graphs the leaf conductance for carbon dioxide is plotted against the net assimilation.

The actual value is for maize about 120 vppm. A similar relation seemed to hold for bean (Phaseolus), a C_3 plant, but here the intercellular CO_2-concentration was about 210 vppm (Goudriaan & van Laar, 1978). Verification experiments, which are discussed in

8.2, suggest that this regulation of the intercellular CO_2-concentration is common for field grown maize and therefore this feature is included in the simulation program.

Then the maximum net assimilation is set equal to

$$F_m = (C_i - C_c)/r_m \tag{5.2}$$

in which r_m is the mesophyll resistance at high light intensity and full turgidity, C_i the intercellular CO_2-concentration and C_c the compensation point for CO_2. The CO_2 compensation point for C_3 and C_4 plants is 50 and 10 vppm, respectively and the intercellular CO_2-concentration is assumed to be regulated around 210 and 120 vppm, respectively. This value may vary over a 20% range. Of course, internal CO_2-concentrations cannot be regulated at these values if the external CO_2-concentration is too low. Based on estimates of Goudriaan & van Laar, the actual setpoints are taken as the minimum of 120 and 0.4 times the external CO_2-concentration or 210 and 0.7 times this concentration for C_4 and C_3 plants, respectively.

Sinclair *et al.* (1977) suggested a procedure to calculate the mesophyll resistance at high light intensities from leaf thickness, cell size, content of carboxylating enzymes and so on, but this procedure has not yet been incorporated in the program. Especially the mesophyll resistance is assumed to be a function of temperature.

The resistance to CO_2 diffusion from the external source towards the intercellular space is now calculated with the equation:

$$\sum r = (C_e - C_i)/F_n \tag{5.3}$$

In which C_e is the external CO_2-concentration and F_n is the net assimilation calculated according to Eqn (5.1). The stomatal opening, calculated from $\sum r$ with $r_l = \sum r - r_b$, in which r_b is the resistance to CO_2 diffusion in the boundary layer, is therefore a result of the rate of CO_2 assimilation which controls in this way the rate of transpiration. The water status of the crop may, however, also determine the rate of transpiration since the minimum possible stomatal resistance increases with decreasing relative water content. Care should be taken therefore that the minimal stomatal resistance at full turgidity is lower than the resistance calculated with Eqn (5.3); this adjustment problem is further treated during evaluation in 8.2.

Eqn (5.2) implies that the maximum assimilation rate depends on the intercellular CO_2-concentration. For C_3 plants, the initial efficiency also depends on this concentration, and is then, according to

Charles-Edwards & Ludwig (1974) best described by a saturation function of the type $y = bx/(x + a)$ in which y is the efficiency, x is the CO_2 concentration and a and b are constants, fitted to experimental data. The value of a may be of the order of 200 vppm CO_2 (Ehleringer & Björkman, 1976). These dependencies on the CO_2 concentration are of minor concern when the intercellular concentration is regulated as in Fig. 15b, but should be considered in situations where this regulation is absent, as in Fig. 15a.

The present programs include full regulation but when necessary can be adapted for no regulation or intermediate situations where the intercellular CO_2-concentration is regulated at a value that depends on the external concentration.

The difference between external and internal CO_2-concentration of C_3 plants is about $300 - 200 = 100$ and of C_4 plants is $300 - 100 = 200$ and this difference, rather than the difference in CO_2 compensation point explains why the transpiration coefficient of C_3 plants is about twice that of C_4 plants at any light intensity and at normal CO_2-concentrations. At low light intensity, the assimilation of the two species is about the same (Fig. 12), but the stomata of the C_4 species are so much more closed that the difference in internal CO_2-concentration is maintained. The transpiration of C_4 species is then about half that of C_3 species. (Alberda & de Wit, 1961). At high light intensity, assimilation of the C_4 species is about twice that of C_3 species (Fig. 12) so that the stomatal conductances are the same to maintain the difference in internal CO_2-concentration. Then the transpiration rate is also about the same.

When the stomatal aperture is determined by the water stress in the leaf, it governs the intercellular CO_2-concentration and with this the assimilation. The intercellular CO_2-concentration is then lower than without water stress and therefore the transpiration coefficient is somewhat smaller; this phenomenon has indeed been observed by Lof (1976). Ageing of leaves is reflected in an increase of the mesophyll resistance. This leads to lower assimilation, and consequently to a lower stomatal aperture and a reduction in transpiration. Thus these results imply that stomatal closure upon ageing is the result of lower assimilation and not its cause so that the transpiration coefficient is hardly age dependent.

Finally it is emphasized that any regulation of the stomatal aperture leads to partially closed stomata and that breeding for plants without such regulation would result in higher intercellular CO_2-concentration, higher assimilation of CO_2 and higher transpiration.

6 Plant synthesis, respiration and reserve utilization

The relational diagram of Fig. 1 illustrates that photosynthesis products of the leaves are supposed to enter a reserve pool and that these reserves are utilized for the growth of structural dry matter and respiration. Although the photosynthesizing leaves produce a mixture of carbohydrates, amino acids and organic anions, only the carbon balance is treated here in more detail, it being assumed for reasons of simplicity that all carbon reserves are weighted as starch.

The CO_2 evolution caused by growth processes is called growth respiration. Maintenance processes counteract the continuous degradation of proteins and ion concentrations. The CO_2 evolution associated with it is called maintenance respiration. This chapter considers the nature of growth and maintenance processes, their efficiencies and the concurrent respiration. It describes also how rates of shoot and root growth are calculated.

The processes of synthesis and maintenance have been treated in considerable detail by Penning de Vries and coworkers (1972; 1974; 1975a,b; 1977a,b). It suffices therefore to summarize only those results of these papers that are relevant for the present more primitive approach. The programming aspects of growth, maintenance and respiration are described in Chapter 7.

6.1 Principles

The underlying assumption of this section is that biochemical reactions are the basis of growth, and that quantification of the reaction equations represents in a realistic way growth processes under most field and laboratory conditions.

For instance, the synthesis of the amino acid ornithine may be represented by the reaction equation

$$\begin{aligned}&1\ \text{glucose} + 2\ NH_3 + 1\ \text{ATP} + 1\ \text{NAD} \rightarrow \\ &\quad 1\ \text{ornithine} + 1\ CO_2 + 2\ H_2O + 1\ \text{NADH}_2 + 1\ \text{ADP} + 1\ P_i \qquad (6.1)\end{aligned}$$

One molecule of glucose and two molecules of ammonia are combined into one ornithine molecule and one CO_2 molecule is released. Some energy is consumed, and this is provided by breakdown of ATP into ADP and P_i. On the other hand, reducing power is retained by reducing the coenzyme NAD to $NADH_2$. ADP and P_i are recombined to ATP at the expense of glucose. The hydrogen in $NADH_2$ may cause formation of 3 ATP molecules by its mitochondrial oxidation, but it may also be used in other synthesis reactions. Hence, during the formation of 1 ornithine molecule, energy is released of which the equivalent of 2 ATP molecules is saved for other uses. This may for example be consumed in the formation of a protein, where presumably 3 ATP molecules are required to provide the energy for joining 1 amino acid to a protein string.

Eqn (6.1) cannot be used as such, since it includes intermediates like ATP and $NADH_2$. To cancel these, it is necessary to account for the cost of their synthesis or the result of their degradation. Synthesis of ornithine protein, expressed in grammes rather than in grammolecules may then be presented by

$$184.7\ \text{g glucose} + 34.0\ \text{g}\ NH_3 + 21.0\ \text{g}\ O_2 \rightarrow$$
$$132.0\ \text{g ornithine protein} + 50.9\ \text{g}\ CO_2 + 56.8\ \text{g}\ H_2O \quad (6.2)$$

A detailed program to calculate such equations for the majority of plant components and their combinations was developed by Penning de Vries, *et al.* (1974). It is known that the relative abundance of amino acids in proteins varies a great deal between types of proteins. However, by applying the detailed program for different types of proteins, it was found that the ratio of the weight of the substrate used and the protein formed varied only marginally, so that all proteins can be lumped together. The same holds within the group of fatty acids (lipids), organic anions and structural carbohydrates. The lignin-like substances are also characterized as one group, too little being known about their biochemistry. Also it should be noted that the detailed program assumes a highly efficient internal use of energy and carbon.

The characteristic values of conversion of glucose into categories of plant substances are summarized in Table 2. The group organic N-compounds consists of 87% proteins, 10% free amino acids and 3% nucleic acids. The 'production value' in this table represents the weight of the end product formed from 1 g of glucose. The 'oxygen requirement factor' and the 'carbon dioxide production factor' are the amount of oxygen consumed and the amount of CO_2 produced during the conversion of one gram substrate, respectively.

The weight of organic N-compounds formed with N and S in the

Table 2. Conversion characteristics for synthesis of five categories of plant substances from glucose

	production value $g\,g^{-1}$	CO_2 production factor $g\,g^{-1}$	oxygen requirement factor $g\,g^{-1}$
Carbohydrates	0.86	0.07	0.051
Lipids	0.36	0.47	0.035
Lignin	0.46	0.27	0.090
Organic acids	1.43	−0.25	0.13
Organic N-compounds with NO_3	0.47	0.58	0.030
Organic N-compounds with NH_3	0.70	0.15	0.74

form of NO_3 and SO_4 is much lower than with N and S in the form of NH_3 and H_2S because of the energy requirement to reduce the oxidized forms. The production value for various carbohydrates is close to 1, because most glucose molecules undergo little modification before their incorporation into structures. The production value of glucose for lipids (fats) is only 0.36, largely because much oxygen is removed from the C skeletons.

Provided that the chemical composition of the plant is known, the data of Table 2 enable the calculation of the amount of plant material that may be synthesized from 1 gram of glucose, and the concurrent CO_2 production and O_2 consumption. Then about 1 gram of glucose is necessary for the active uptake of 30 grams of minerals, and the energy for the loading and unloading of the phloem with 1 gram of glucose is derived from about 0.035 grams of glucose. In this way one can calculate that about 0.7 gram of a young maize plant can be formed from 1 g of glucose, NO_3 and the necessary minerals. This result agrees reasonably well with data obtained from maize embryos growing on glucose, nitrate and minerals (Penning de Vries, 1972); results calculated for germination of bean and groundnut seedlings from their reserves are also in good agreement with experimental results (Penning de Vries & van Laar, 1977a).

6.2 Growth

The principles outlined above are used to calculate conversion efficiencies in crops. However, the situation is here considerably complicated because part of the growth processes considered in 5.2 and part of the maintenance processes occur in autotrophic tissue. Many complications are circumvented by assuming that photosynthesizing leaves do not grow and growing organs do not photosynthesize. This is a fair assumption for a plant like maize, where leaf growth takes place at the base within the apparent stem formed by the leaf sheaths. However, in plants like sunflower, growing leaves photosynthesize and here also a fraction of the energy needed for growth may be derived directly from ATP produced by photosynthesis.

As has been said already, it is assumed that all carbon reserves are weighted as starch, although the reserves are delivered to the heterotrophic organs in the form of soluble carbohydrates, amino acids and organic anions. This assumption simplifies the computations considerably, and is at the present stage justified because too little is known about the photosynthetic processes in the leaves, except for CO_2 assimilation. Taking into account conversion and translocation costs in the heterotrophic organs, one can calculate the starch utilization and the concurrent CO_2-evolution with the data of Table 3. Since the reserves are weighted as starch, but amino acids are produced in the photosynthesizing leaves, it suffices to calculate only the costs for formation of the amino acid skeletons. This holds also for the organic anions which are assumed to be formed in the photosynthesizing leaves during NO_3^- reduction according to the general reaction:

$$RH + CO_2 + NO_3^- + 8\,H \rightarrow NH_3 + 3\,H_2O + RCOO^- \qquad (6.3)$$

Except in conditions of nitrogen deficiency, the organic anion content of most plants fed on NO_3 is equal to or lower than the content of nitrogen incorporated in amino acids and proteins; the excess organic anions are transferred to the roots and decarboxylated (Dijkshoorn, 1971). This decarboxylation results in an additional evolution of CO_2 in the roots and the formation of glucose. All translocation costs of carbohydrates are incorporated in the synthesis costs of carbohydrates, lipids, lignin and protein. Consequently, one should not take into account again the costs of translocating this glucose in the form of organic anions in the heterotrophic tissue.

Although part of the organic anions that are formed during nitrate reduction may remain where they are formed, the assumption of spatial separation between photosynthesis and growth necessitates that one includes the translocation costs of all organic anions

Table 3. Basic data for heterotrophic growth

conversion only		
1.25 g starch	→	the C-skeleton for the amino-acids needed for 1 g proteins
0.892 g starch	→	the C-skeleton for 1 g organic anions
0.0391 g starch	→	energy for the translocation of 1 g organic anions in heterotrophic tissue
0.035 g starch	→	energy for the uptake of 1 g of an average mixture of minerals and NO_3
1 g starch	→	1.63 g CO_2 upon respiration
0.366 g starch	→	C lost by CO_2 evolution during decarboxylation of 1 g organic acids
conversion plus translocation		
1.12 g starch	→	1 g carbohydrates + 0.175 g CO_2
2.73 g starch	→	1 g lipids + 1.618 g CO_2
1.94 g starch	→	1 g lignin + 0.620 g CO_2
1.25 g amino acids + 0.518 g starch	→	1 g protein + 0.844 g CO_2
0.018 g starch	→	1 g translocated minerals + 0.03 g CO_2

that remain in the plants. It should be realized that any further sophistication will only lead to corrections of a minor respiratory term. For the same reason, sophistication in the treatment of mineral uptake and translocation in the tissue is not necessary.

To apply the information of Table 3, the composition of the daily weight increment has to be known. The simplest assumption is that the plants maintain the same composition so that the increment composition does not vary with time. Another way of programming increment composition may be based on the assumption that the composition of optimally fertilized plants depends on the physiological age, but it should be realized that this oversimplifies the influence of environmental factors. In the present model a more cautious way of programming is chosen. Increment composition is introduced by a forcing function based on the data collected from the verification experiments to be simulated.

6.3 Maintenance

Even at minimum carbohydrate levels, energy consuming processes exist that are necessary to maintain the plant structure. These maintenance processes compensate for the degradation of existing structures at a cellular level of organization, for instance, resynthesis of hydrolysed proteins. Maintenance can thus be treated as growth that counteracts degradation, so that the corresponding use of starch can be calculated from the turnover rates of the cell constituents and their cost of formation from the remaining building blocks. However, it is difficult to obtain reliable data on life expectancy of chemical constituents and ion gradients, as was experienced by Penning de Vries *et al.* (1974), who evaluated the biochemical information.

The average turnover rate of the various leaf proteins may be about 100 mg proteins per gram protein per day or 0.1 day^{-1} at normal temperatures in leaves assimilating at moderate light intensities. Their resynthesis requires about 0.24–0.43 mg glucose per g protein which equals about 7–13 mg glucose per g dry weight per day in leaves.

The turnover rate of membranes is about 1 day^{-1}. From total membrane weight and an assumed fraction of proteins and lipids that are completely degraded, the cost of membrane maintenance is estimated at 1.5 mg glucose per gram dry matter per day for tissues with a normal N-content.

The turnover rate of cell walls is so small that the cost of their maintenance is negligible. The energy cost of maintaining potential gradients of ions between the cytoplasm and its environment is considerable. From a rough estimate of active fluxes through plasmalemma and tonoplasts and from the assumption that the energy of 1 ATP is needed for each active transfer, it may be concluded that the total energy requirement for maintenance of ion gradients amounts to 6–10 mg glucose per g dry matter per day.

There are no indications that a noticeable amount of energy is needed for maintaining gradients other than those of ions, for providing heat or for movements of organs or protoplasm. Although it has been suggested that respiration without any useful purpose may occur ('idling' or 'uncoupled' respiration), there is so little evidence for this that it has been neglected.

In the model, maintenance respiration is not directly related to dry weight but to the actual amounts of protein and minerals in the plant. In this way, changes in plant composition are reflected in the maintenance rate. For the shoot, one should remember that part of the tissue consists of photosynthesizing leaves and that their maintenance is already accounted for. Hence, for maintenance calculations the shoot weight should be reduced by this weight, which is equal to the leaf area index times the specific leaf weight, the latter being set for maize at 750 kg ha^{-1} (leaf) if detailed information is lacking. The total rate of maintenance respiration, calculated from these basic data, is about 0.0225 g starch per gram protein per day and 0.03 g starch per gram minerals per day and is about correct for plants grown under moderate or low light intensities. It is too low for plants growing in higher light intensities, probably because the protein turnover rate is underestimated. This is corrected by the addition of a term which characterizes the metabolic activity and which is assumed to be proportional to a first order average with a time constant of half a day of the starch use in the plant. The numerical value is assumed to be 0.04 g starch per gram of starch used. This comes close to a straightforward fudge factor and the need for its use illustrates that in a final analysis it may be difficult to maintain the present complete separation between growth respiration and maintenance respiration.

The estimates hold at normal temperatures of 20–25°C. Since protein turnover and ion fluxes are considerably temperature dependent, maintenance respiration is also affected by temperature. It is not clear in what range the classical 2–3 fold increase in maintenance respiration with a 10°C increase in temperature occurs because most measurements of maintenance rate are obscured by growth respiration, but this Q_{10} of 2–3 is applied over the whole temperature range. Any direct effect of temperature on the dark respiration of the photosynthesizing leaves is omitted because of the assumption that maintenance contributes only little to its magnitude.

It appears from the whole treatment that there is indeed scope for further research on the quantitative aspects of maintenance, and especially as this process may affect the net production rate of crops to a considerable extent. After all, a difference in 10 mg starch g dry $matter^{-1}$ day^{-1} means 100 kg starch ha^{-1} day^{-1} for a standing crop of 10 000 kg dry matter ha^{-1}.

6.4 CO_2 dissimilation

By applying the constants for CO_2 evolution in Table 3, the rates of CO_2 dissimilation can be directly derived from the starch utilization and conversion processes. The CO_2 evolution due to mineral uptake and that due to decarboxylation of organic anions contribute to root respiration. They must be distinguished because during daily measurements of CO_2 assimilation the root respiration does not interfere with the changes in CO_2 concentration within the enclosure.

The programming is tedious and mistakes can be easily made, therefore it is advisable to check the carbon balance independently. The amount of carbon gained by assimilation minus that lost by shoot and root respiration plus the amount taken from the reserves is compared with the amount of carbon that is incorporated in the various chemical consitituents. In the present program, both sides of the equality differ less than 1 percent, a deviation which may be attributed to rounding-off errors. This agreement indicates that no obvious programming errors are made, and that the sum of the chemical fractions of the plant adds up to one. It does not give an indication about the acceptability of the underlying assumptions.

6.5 Growth rates of root and shoot

The approach to relate the rate of carbohydrate consumption to the rate of growth of structural material has been discussed, but the problem of the magnitude of these processes remains. The growth rates partly depend on the amount of tissue capable of growth, the temperature, the relative water content and the amounts of carbohydrate reserves that are available.

Penning de Vries *et al.* (1979) paid considerable attention to the relation between respiration rate and the concentration of total soluble carbohydrates in plants, assuming that the first reflects growth and the second the availability of reserves. Results for wheat, ryegrass and maize are presented in Fig. 16. In spite of the unavoidable scatter in the observations, first the respiration increases more or less linearly with the concentration of the total soluble carbohydrates, but the response may level off at concentrations of 10–20%.

For species that actually accumulate starch, the total soluble carbohydrates are only part of the reserves so that the problem becomes more complicated as shown by data of Challa (1976) for cucumber. The relation between respiration and total soluble carbohydrates was again linear for so-called winter plants (low light intensity, short days) whereas a saturation level was observed for spring plants (Fig. 17). Challa also observed that the rate of conversion of starch into sugars was proportional to the amount of starch

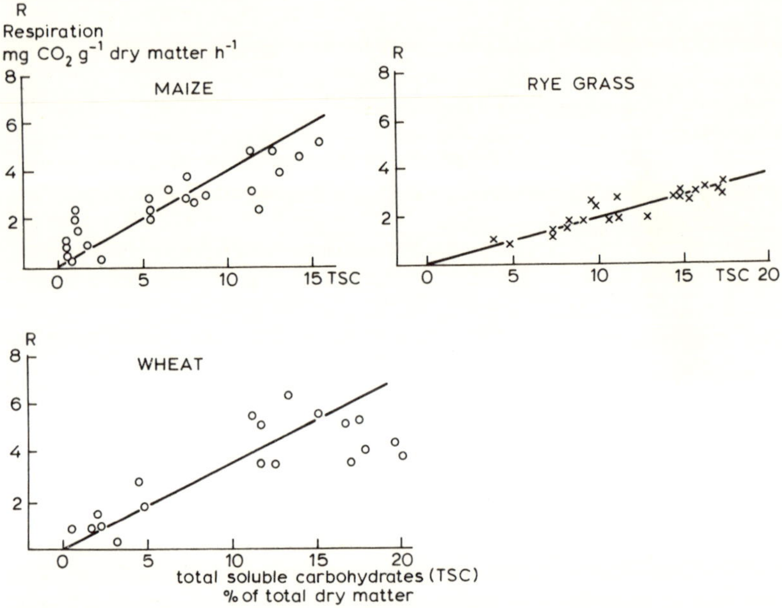

Fig. 16. The relation between respiration rate and the amount of total soluble carbohydrates in the plants measured for some period after the onset of darkness.

accumulated during the day (Fig. 18). The result is that in each situation an equilibrium level of sugars is maintained in the plant at which the rate of sugar consumption due to growth and respiration equals the rate of production from starch. The sugar level decreases upon rate of depletion of the starch store.

Since starch accumulation is not very evident in the species that are at present under consideration and since levels of carbohydrate reserves higher than 20% do not occur, it is assumed here that the consumption rate of carbohydrate reserves for growth is linearly related to the reserve percentage. Therefore the rate of growth may be set proportional to the total amount of reserves in the canopy by a relative consumption rate of reserves. This relative rate must be assumed to be about 1 day^{-1} under optimal conditions to reflect the observation that the daily production of reserves is then used for growth without accumulation at a high level.

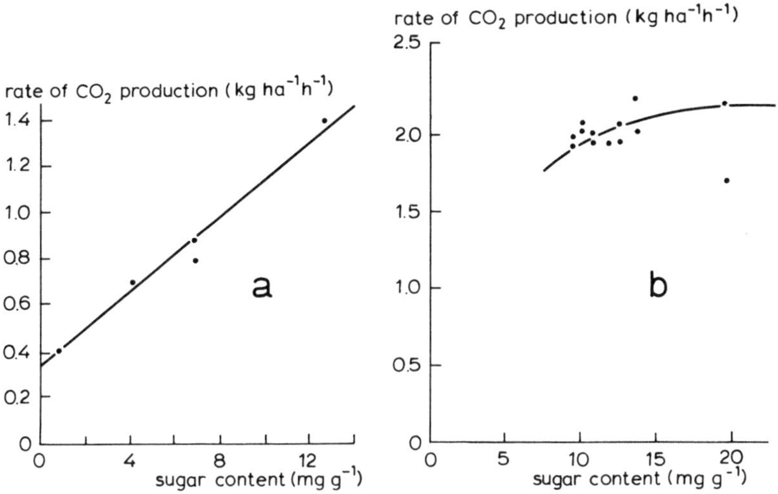

Fig. 17. Relation between sugar content in the leaves and the rate of CO_2 production of the shoot: (a) winter plants, (b) spring plants (Challa, 1976).

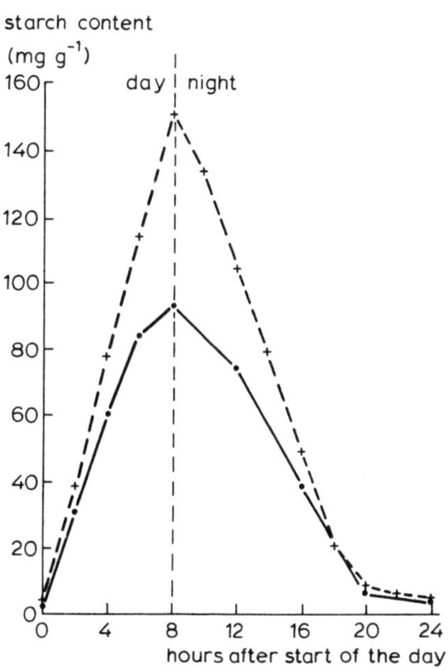

Fig. 18. Diurnal course of leaf starch content of winter plants, cultivated under a raised CO_2-concentration of 1700 vppm (+ – – – +), compared with the standard winter plant (·———·). Contents are expressed on basis of starch and sugar free dry weight (Challa, 1976).

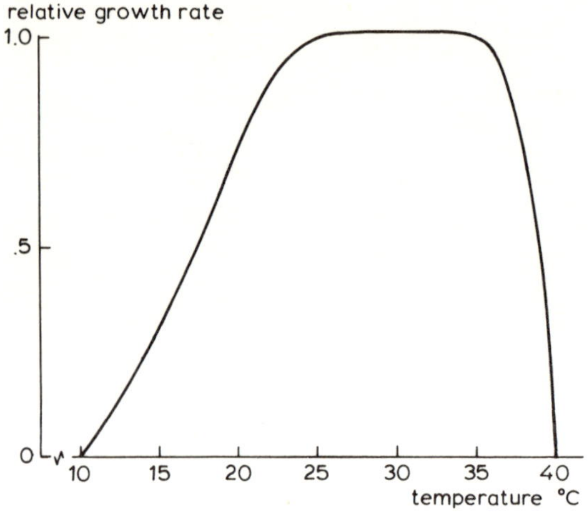

Fig. 19. Relation between relative growth rate and temperature.

The multiplication factor to account for the influence of temperature on the relative consumption rate of reserves is given in Fig. 19 for maize, as derived from de Wit *et al.* (1970). This curve is again assumed to hold for plants pretreated at 25°C. Temperature adaptation of growth is also obvious and could be accounted for in the same way as for assimilation in Fig. 13.

It should be noted that the influence of temperature on growth is independent of that on assimilation. Thus at low temperatures, a situation may occur where assimilation is so much higher than growth that the reserves accumulate to unrealistically high levels. This may be avoided by assuming that the relative effect of temperature on growth and assimilation is about the same, but most evidence points to the contrary. Unrealistic accumulation of reserves may also be avoided by assuming that their level feeds back on assimilation, for which there is reasonable evidence, although the physiological mechanism remains unclear (Neales & Incoll, 1968).

In the model this assumption is introduced by dividing the mesophyll resistance by a factor which is 1, for reserve percentages lower than 20, and practically zero for percentages higher than 25, two rather arbitrarily chosen boundaries.

The consumption of reserves is partitioned over root and shoot according to a distribution function dependent on the relative water

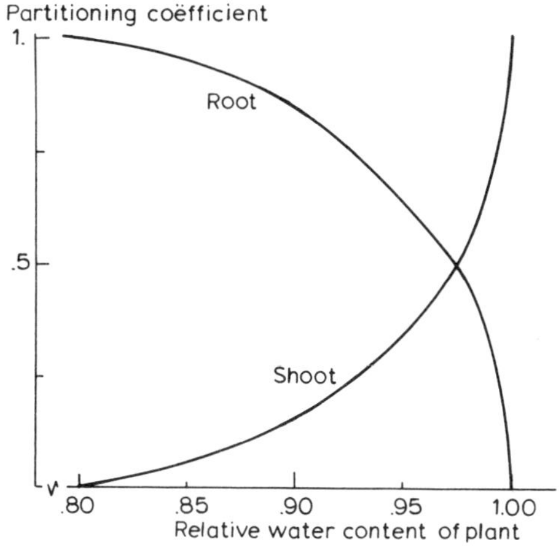

Fig. 20. Distribution of dry matter over root and shoot in relation to the relative water content of the plant.

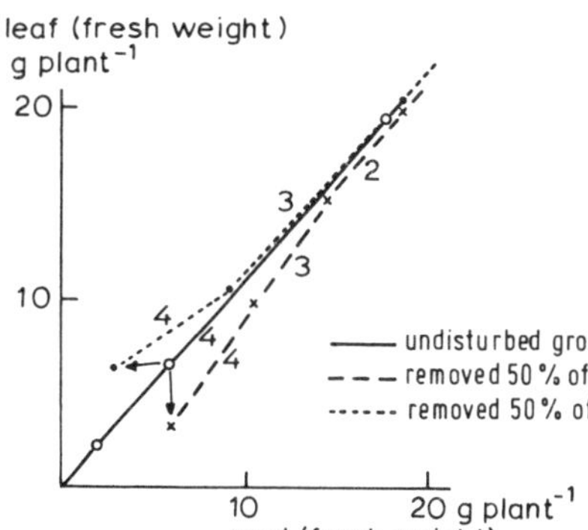

Fig. 21. Recovery of the original leaf-root ratio of bean plants (O——O), after removal of parts of the leaves (× -- ×) and of the roots (.......) (Brouwer, 1963). The numbers indicate the period between measurements in days.

content of the crop, as illustrated in Fig. 20. Root growth is promoted at low relative water contents and shoot growth at higher ones. This procedure ensures that a functional balance between root and shoot growth is maintained.

A sudden disturbance of the shoot-root ratio is followed by a recovery towards the equilibrium level as shown in Fig. 21. The time period needed to reach a new equilibrium is of the order of 5–10 days. This time span may also be arrived at analytically by applying the concept of the time constant (de Wit & Goudriaan, 1972). Here the definition of the time constant is:

$$T_a = SRR \Big/ \frac{dSRR}{dt} \tag{6.4}$$

where *SRR* is the shoot-root ratio.

Elaboration of this expression yields

$$T_a = \frac{S \cdot R}{R(dS/dt)S(dR/dt)} \tag{6.5}$$

in which

S = shoot weight
R = root weight

The growth rates of shoot and root are given by

$$dS/dt = GR \cdot F \tag{6.6}$$

$$dR/dt = GR \cdot (1-F) \tag{6.7}$$

GR is the growth rate of the whole plant, and *F*, between 0 and 1, is the partitioning coefficient. In its turn *GR* equals the total plant weight divided by the inverse of the relative growth rate or the time constant of growth, that is $(S+R)/T_g$. Hence

$$dS/dt = ((S+R)/T_g) \cdot F \tag{6.8}$$

$$dR/dt = ((S+R)/T_g) \cdot (1-F) \tag{6.9}$$

Substitution of these equations in Eqn (6.5) yields

$$T_a = \frac{S \cdot R}{S+R} \cdot \frac{1}{R \cdot F - S(1-F)} \cdot T_g \tag{6.10}$$

After root removal, the relative water content decreases and *F* approaches zero, so that

$$T_a \simeq -\frac{R}{S+R} \cdot T_g \tag{6.11}$$

and for shoot removal *F* approaches 1 so that

$$T_a \simeq + \frac{S}{S+R} \cdot T_g \qquad (6.12)$$

The opposite signs confirm the existence of a functional balance. At a reserve level of 15 percent and a relative consumption rate of reserves of 1 day^{-1}, the relative growth rate of the plant is 0.1 day^{-1}, so that T_g is 10 days. With a relative shortage of roots, $R/(S+R)$ may be around $\frac{1}{5}$, so that the time constant of adaptation is of the order of 2 days. With a relative shortage of shoots, $S/(S+R)$ may be about 0.5, so that the time constant of adaptation is about 5 days, or even longer because CO_2 assimilation is decreased by shoot removal and consequently the time constant of growth (T_g) is increased.

The order of magnitude of the time constant and especially the difference in value between the time constants for removal of root and shoot are in good agreement with observational results, an example being given in Fig. 21 (Brouwer, 1963).

The assumption of complementary growth governs also the shoot-root ratio dependence of the evaporative demand. In case of equilibrium

$$\frac{dR}{dt} \Big/ R = \frac{dS}{dt} \Big/ S \qquad (6.13)$$

so that with Eqns (6.6) and (6.7) it follows that

$$R/S = (1-F)/F \qquad (6.14)$$

At equilibrium, transpiration equals uptake. The difference in water potential between shoot and root medium is proportional to $(1-F)$ and so is the uptake. Assuming stomatal regulation by water stress, the transpiration is about proportional to F and with the evaporative demand, E. Hence

$$R(1-F) = K \cdot S \cdot F \cdot E \quad \text{or} \quad (1-F)/F = K(S/R) \cdot E \qquad (6.15)$$

Substitution of this expression in the previous equation yields

$$R/S = \sqrt{K \cdot E} \qquad (6.16)$$

The strict linearities assumed in this derivation do not occur in the simulation program, so that the above square root dependency of the shoot-root ratio on evaporative demand is approximate. However, the derivation shows that the simulation program contains the necessary feedbacks to guarantee a realistic relation between shoot-root ratio and evaporative demand.

7 Description of the simulation models

In Chapter 2 the system to be simulated was described qualitatively and a procedure for verification was outlined. In Chapters 3–6 the main components of the system were treated in detail, the theoretical background of the different processes was discussed and quantitative relations between the driving variables and the various rates were described. In these chapters the many gaps in our knowledge were revealed by the various estimates and short cuts that were necessary for the construction of the crop growth model.

In this chapter two versions of the actual computer model are presented in such a way that together with the comments in the print-out (Appendices A and B), outsiders will be able to use them. The first version (BACROS) simulates the growth of a crop over a whole growing season, while the second version (PHOTON) simulates the processes of photosynthesis, respiration and transpiration during one day. The differences are that in BACROS provision has been made for time intervals of one hour (as treated in Chapter 4) while PHOTON contains options for the comparison of the simulated results with measurements in the crop enclosure in the field. Appendix C gives the dimensions used in the model and an explanation of the abbreviations.

7.1 The basic crop growth simulator (BACROS)

A listing of the model for seasonal growth is given in Appendix A. The model is divided into 10 sections.

Section 1 of the program contains the MACRO definitions, which have to be defined before the structure of the model. A so-called MACRO is a functional block containing basic functions that may be used several times in a program. Each time a MACRO is called upon in the model the set of statements contained in its definition is inserted at that particular place. During this operation the 'dummy' input and output variables defined in the MACRO label card are replaced by their equivalents in the call statement.

The first MACRO describes the calculation of the sensible and latent heat loss, the leaf temperature and the photosynthesis per unit of leaf area. This MACRO is called on several times in the energy balance section, its use being discussed there in detail. MACRO 2 contains a function which calculates daily totals for various rates of change, such as water loss, growth of plant constituents and dry matter increase. In MACRO 3 the chemical composition of the plant material growing at any particular moment is calculated. As already explained in Chapter 6, this composition is given here as a forcing function by introducing the measured composition at discrete time intervals. In MACRO 4 a semi-sinusoidal daily course is calculated for minimum, maximum and dew point temperatures. The form of the function depends on the time of sunrise and sunset and the actual values of the measured daily minimum, maximum and dew point temperatures, as explained in Chapter 3.

Section 2 deals with initialization of the model. First the geometrical parameters pertaining to leaves and radiation are calculated in a procedure, which is described in detail by Goudriaan (1977). Next the variables defining the site of the experiment are given and finally the state of the crop at the onset of the simulation is described by the initial values of the relevant state variables. The quantitative aspects of this initialization are treated in more detail in Chapter 8.

Section 3, the first section of the dynamic part of the model, treats the weather conditions. Subsection 3.1 contains the calculation of the position of the sun in the sky, air and soil temperature, wind speed and the humidity of the air. In Subsection 3.2 the radiation climate is calculated, for both the diffuse and the direct components, and the extinction coefficients for both types of radiation under the assumption of exponential extinction. A detailed treatment of the processes defined in Section 3 is given by Goudriaan (1977).

Section 4 is the backbone of this simulation model. It contains the calculation of the energy balance and the photosynthetic performance of the leaves. These calculations are repeatedly executed for subsequent leaf layers, each containing a total leaf area index of one. Within each leaf layer the absorbed and reflected radiation in the different wavelength regions, visible, near-infrared and long wave are first calculated applying the methods given in Chapter 2. A separate provision is included for calculations at night time, when the incoming global radiation is zero so that a large part of the calculations can be skipped.

Wind speed within each layer is obtained next, assuming exponential extinction with depth in the canopy. This wind speed is applied in a semi-empirical formula to calculate the resistance to exchange of water vapour and CO_2 in the boundary layer of the leaves. With this resistance and the vapour pressure deficit of the air, the 'drying power' term of the transpiration is calculated.

Now the MACRO in which the transpiration and assimilation of individual leaf layers is calculated can be applied, the necessary inputs being: the radiation intensity of visible, near-infrared and long-wave radiation and the differential leaf area index for each class of leaves. Within each leaf layer, the MACRO is applied for overcast and for clear skies. In the latter case a distinction is made between leaves directly exposed to the sun and shaded leaves. Provisions are made for the total canopy leaf area index being smaller than 0.2 and for the leaf area index of one leaf layer being smaller than unity. Within the MACRO, the net assimilation rate of individual leaves is obtained by interpolation in the CO_2 assimilation curve, described by Eqn (5.1). This net assimilation rate is applied in a Penman-type equation (Chapter 4) to calculate the stomatal resistance of the leaves. Next it is checked whether the resistance calculated in this way, is less than that expected from the water status of the leaves. If so, the net assimilation is recalculated, using the stomatal resistance obtained from the relative water content of the leaves. The total leaf resistance for water vapour exchange is obtained by assuming that transport takes place through the stomata until they are completely closed, after which only transport through the cuticle is considered. The evaporative heat loss of the leaves is calculated by applying again a Penman-type formula (Chapter 4), taking into account the energy used for the photosynthetic processes. This energy in $J\,m^{-2}\,s^{-1}$ is obtained by multiplying the net assimilation rate in $kg\,CO_2\,ha^{-1}\,h^{-1}$ by 0.3.

The sensible heat loss is found now by subtracting the evaporative heat loss and the photosynthetic energy consumption from the total absorbed radiation. Leaf temperature follows from the sensible heat loss through application of Ohm's law to the difference between air and leaf temperature. The heat exchange resistance in $°C\,m^2\,s$ $Joule^{-1}$ equals the resistance of the laminar layer in $s\,m^{-1}$ divided by the volumetric heat capacity of the air. The total evaporative and sensible heat loss and the net assimilation of carbon dioxide per leaf layer is obtained now by multiplying the rates for the individual leaves by the surface area considered. Whole canopy totals for heat loss and net assimilation are finally found by multiplying the rates calculated for clear and overcast skies by the fractions of time that these conditions exist.

Section 5 deals with the water balance of the canopy. The

iteration procedure, necessary for application of time intervals of one hour in the simulation is described. An arbitrarily chosen error criterion on the difference between water uptake and transpiration, with a relative value of 0.005 or an absolute difference of 0.00004 $g\,m^{-2}\,s^{-1}$, results under normal conditions in an equilibrium situation after 3 to 4 iterations (Chapter 4). The iteration loop is constructed in such a way that only those statements of the simulation program that are needed for the iteration are repeatedly executed. This is achieved by a careful choice of the variables included in the relevant procedure statements.

Section 6 contains the variables associated with the water status of the canopy. A detailed description is not necessary here as this point was extensively treated in Chapter 4, and with the comments in the print-out this section is self-explanatory.

In Section 7 the processes connected with reserve accumulation are treated. First photosynthetic processes are considered. The parameters describing the net assimilation curves, used in Section 4, are obtained. The maximum assimilation rate in kg CO_2 $ha^{-1}\,h^{-1}$ is calculated from the mesophyll resistance, and the difference between the internal CO_2-concentration and the CO_2 compensation point. The internal CO_2-concentration is calculated as a fraction of the external concentration but has a set maximum value. The numerical values of the parameters used may depend on plant species. Also the dissimilation rate in the photosynthesizing tissue is defined in this section, and is dependent on an average net assimilation rate. This section contains also a description of a procedure to account for the effect of temperature adaptation on CO_2 assimilation.

In Subsection 7.2 the reserves of the plant are considered. The amount of reserves increases due to CO_2 assimilation and decreases due to growth and maintenance processes. The rates of increase and decrease were discussed in Chapters 5 and 6, respectively. Standardization is achieved by expressing the reserves in amount of starch. To obtain the rate of increase of reserves from the assimilation rate, a weight ratio of 1.63 kg starch per kg CO_2 is introduced (molecular weight ratio per unit carbon).

The level of reserves decreases by starch consumption due to growth and maintenance processes in shoot and root. The efficiency of growth (Subsection 7.3) is obtained by multiplying the growth rates of the various plant components in root and shoot by the starch requirement factors, given in Table 3. The starch required for the formation of the C-skeletons of amino acids and organic anions is also taken into account, as is the carbon lost during decarboxylation of the excess organic anions in the roots.

In Subsection 7.4 the maintenance costs of the tissue are calculated for the fraction non-assimilating tissue of the shoot and for the root, it being assumed (6.3) that in the autotrophic tissue these costs have been accounted for in the curve of measured CO_2 assimilation. In this process resynthesizing hydrolysed proteins and maintaining ionic gradients are taken into account by considering the protein and mineral fractions of the material. The difficulties arising from the schematized distinction between growth and maintenance processes are circumvented by the introduction of costs to maintain an average metabolic activity in shoot and root (6.3). The temperature dependency of the maintenance processes is introduced by a multiplication factor, expressing the classical 2-fold increase per 10°C temperature difference. In the starch requirement for the root, the energy needed for the uptake of nitrate and other minerals is taken into account.

In Subsection 7.5, the CO_2 evolution connected with growth and maintenance processes is calculated for both, shoot and root. The CO_2 production factors used are those derived by Penning de Vries *et al.* (1974). They are given in Table 3.

In Subsection 7.6, a carbon balance is calculated, which is used to check the program for mistakes. Such balance equations are a useful way of testing complex models for internal consistency. In the present model the difference between carbon present and the accumulated balance of fixation and evolution should not exceed 1%, small deviations being due to rounding-off errors during the calculation.

Section 8 considers growth of the canopy. In Subsection 8.1, the measured values of shoot and root weight are introduced for comparison with the simulated values. In Subsections 8.2 and 8.3 the growth of shoot and root is considered. Both sections are practically the same, with relevant parameter values substituted at the proper places. First one calculates the rate of increase in structural dry weight of shoot and root, which is dependent on the reserve level, the temperature and the water status of the crop, according to the principles described in Chapter 6. Next this structural material is partitioned among the various plant constituents by applying for each of them the MACRO INCREM, governed by the measured chemical composition of the harvested material.

In the shoot section (Subsection 8.2) the rate of nitrate reduction is calculated under the assumption that all the nitrogen required for protein synthesis is imported as nitrate. The formation of organic anions proceeds concurrently and their rate of transport to the roots is calculated. Sometimes the measured concentration of organic anions requires a rate of formation in excess of the rate of nitrate reduction. This is programmed by importing organic anions from the root.

The root section (Subsection 8.3) contains the calculations on root suberization and decay, which results in a partitioning of the total root system between 'old' and 'young' roots.

Finally in Section 9 input data are given: Subsection 9.1 refers to plant parameters, both physiological parameters and measured data. Subsection 9.2 contains the measured meteorological data: maximum and minimum values of daily temperature and of dew point temperature, wind speed and daily total global radiation. All meteorological data are in principle measured in a Stevenson screen in the field that is simulated. These data are used in Sections 2 and 3 to generate daily courses.

7.2 The computer model for simulation of assimilation, respiration and transpiration throughout the day (PHOTON)

The version of the model (Appendix B) that simulates transpiration, assimilation and respiration throughout the day has basically the same structure as the version for crop growth during the season. The comments in the computer text make detailed discussion of all differences between PHOTON and BACROS unnecessary: it suffices to treat main differences.

1. The list of input data at the beginning of the program concerns physiological parameters, physical properties of the assimilation chamber or crop enclosure and run control.
2. In Section 1 the MACRO on assimilation, transpiration and respiration is given. The MACROs for calculating daily totals, for generating daily courses of meteorological forcing functions and for computing daily increments of the plant materials are not used in this version.
3. In Section 3, the influence of the crop enclosure on the radiation is treated.
4. Sections 4 and 5 are considerably shorter because of the absence of the iteration procedure that allows time intervals of one hour for the water balance of the plant. Here, the time intervals of integration, determined by the Runge Kutta–Simpson procedure, are dependent on the smallest time constant of the model, which is often

the inverse of the relative rate of change of the relative water content of the plant.
5. Section 9 enables a direct comparison of the simulated results with the results of the measurements in the crop enclosure. Subsection 9.1 concerns the actual measurements on incoming and outgoing air composition and the net assimilation and accounts for the size and form of the crop enclosure.

8 Performance of the models

8.1 Introduction

During recent years, considerable attention has been given to the problem of model evaluation. Much of the problems and ideas were discussed in a Workshop in Wageningen on 'the evaluation of simulation models in agriculture and biology' (Penning de Vries, 1977c). It was felt there that the word evaluation is the broadest term for assessing the value of a model. It is used for checking internal consistency and units used in the computer program, for comparison of model output with real world data and for assessment of practicability. Validation is then used as evaluation with emphasis on usefulness and relevance of the model and verification for evaluation with emphasis on truthfulness.

The model presented in this monograph is an attempt to explain the behaviour of crop surfaces from the knowledge of the underlying physical, chemical and physiological processes. Evaluation is then done on two levels. The first concerns the verification of the description of the underlying processes, as has been done in the previous chapters. Mostly the description is unbalanced: some processes are handled in detail and in a generally acceptable way, whereas other processes are treated rudimentarily, in a more or less ad-hoc fashion.

A scientific aspect of modelling is the exposure of the imbalance of the treatment of various fields of research which in its turn guides the identification of further relevant research areas. But relevant for what purpose? The model presented here was constructed to obtain a quantitative description of important aspects of crop growth with the use of a minimum amount of experimental field data.

This purpose makes validation of the model by comparison of model output with experimental data gathered under field conditions necessary. This validation may result in the falsification of some model components or may show the necessity of using forcing functions which cannot be conveniently determined in the field situation. In both cases, the validation process may indicate also areas for further study. It should be realized, however, that a model as a whole cannot be falsified, although one may come to the conclusion that the result does not justify the effort. In this respect, a model resembles a car in use: components may be malfunctioning, but the user is reluctant to discard the vehicle until something better turns up.

During validation, the model's performance is in general improved by calibrating it against the field data at hand. This calibration is essentially a process of curve fitting in which weak or unknown parameters are adapted to reach a reasonable overall agreement between simulated and observed results. It is a dangerous procedure because the number of parameters within simulation models are in general large compared with the amount of experimental field data at hand, so that indiscriminate application of this technique could lead to a near-perfect but meaningless goodness-of-fit.

Wherever applied, calibration should be guided by a sensitivity analysis which is most aptly described as a test on the relative influence of realistic changes in input data and parameters on the relevant output of the model. Then in regions of the model where structure is lacking and many parameters are necessary to describe the processes, the sensitivity of the model's behaviour to each parameter is small and calibration leads to nothing. However, in regions with sufficient structure, based on knowledge of the underlying processes, the number of relevant parameters is relatively small and the sensitivity to changes may be accordingly large. Then calibration may provide valuable estimates of parameters.

It is often advocated to execute the various steps of model evaluation in strict order and to keep parameter estimation, based on experimental knowledge of the underlying processes, separated from the estimation by means of calibration. However, model building and evaluation is a continuous effort which leads necessarily to an iterative approach in which the various phases intertwine.

Nevertheless we have tried here to parameterize at first the model as far as possible by means of physiological data obtained from controlled environments with field crops. In this way, the relevant deviations reveal the dangers of indiscriminate extrapolation from controlled environment data to field conditions. Subsequently, some parameters are adapted in a process of calibration and the adapted model is then validated again, but then in reference to the result of periodic harvest experiments in various parts of the world.

The evaluation of the model in this chapter is restricted to maize. The reader is referred to other publications for evaluation of the performance of the model in various stages of development and sophistication, with respect to other plant species. These are: especially perennial ryegrass (van Keulen *et al.*, 1975), Rhodes grass (Dayan & Dovrat, 1977), natural grassland vegetations (van Keulen, 1975), wheat (van Keulen & Louwerse, 1975), soybean (Sinclair & de Wit, 1976) and rice (van Keulen, 1976).

8.2 Crop enclosures

During the first evaluation cycles, there were appreciable differences between the simulated and measured daily course of net assimilation of maize in enclosures. An analysis of these differences showed that especially the relation between maximum assimilation rate and temperature (Fig. 14) and the relation between stomatal conductance and relative water content are different for plants grown under laboratory conditions and in the field. Apparently, the plants adapt themselves to the ever changing environment in the open air in a way which is not sufficiently explained by the analyses of experiments under controlled conditions. Because of this phenomenon, it was necessary to abandon partly the principle that the basic data should be collected exclusively in the laboratory (de Wit, 1970). Therefore a calibration procedure has to be introduced, which resulted in the following assumptions:

The maximum assimilation rate of leaves of field grown maize is 70 kg CO_2 ha^{-1} h^{-1} at temperatures above 13°C and drops linearly to zero within the range from 13 to 8°C. The stomatal conductance of the leaves is 10^{-4} m s^{-1} for relative water contents of the shoot below 0.95 and increases linearly between 0.95 and 1.00 to 0.0143 m s^{-1} when the stomata do not close because of internal CO_2-regulation. At low CO_2-concentrations, the ratio between external and internal CO_2-concentration is 0.6, rather than 0.4 as observed in 5.3. For all other parameters and functional relations, the data as given in the Chapters 2–6 are maintained. Except in a few cases, all subsequent runs with the simulation program in both the PHOTON and BACROS version are executed with these data.

PHOTON simulates the daily course of net assimilation and transpiration of a crop surface. For evaluation, it is necessary to initialize the simulation program with the proper shoot weight and leaf area and with the proper amounts of young and old roots. The first two characteristics are obtained by harvesting the shoot in the enclosure and determining the stem weight, the leaf weight and the specific leaf weight, i.e. weight per unit leaf area. For experiments with plants with less than 6 leaves, about 15% of the leaf area is enclosed by leaf sheaths which is accounted for by considering only the exposed leaf area in the computation of light interception.

Initialization of the root weight is impossible because proper techniques to determine amounts of young and old roots under field conditions are not available. To circumvent this problem, the root weights are initialized from simulation runs with BACROS up to the moment the plants are covered by the enclosure. Year-to-year differences within a climatic region were so small that it suffices to work with an average relation between shoot weight and weight of young and old roots, which is used for initialization in PHOTON. When management practices deviate considerably from the usual ones, especially with respect to planting density, initialization should be based on knowledge of the particular experiment. In many cases, it suffices also to initialize the amount of reserves in the same way, but occasionally it may be necessary to obtain this initial value by using PHOTON for the 24 hours preceeding the onset of the experimental period.

A special experiment was executed in the enclosure to estimate the relation between relative water content and stomatal closure. For this purpose, net assimilation and transpiration were measured for a few hours during a sunny day, the plants being clipped at their base around 11h30 and kept in their original position by a wire construction. At the end of the experiment, the plants were harvested, their fresh and dry weight being determined immediately. From the transpiration data, it was then possible to calculate the water content throughout the experimental period. The experimental results are given in Fig. 22a for the net assimilation and in Fig. 22b for the transpiration. The simulated curves are obtained by assuming that the stomatal conductance decreases linearly from 0.0143 to 10^{-4} m s^{-1} in the range of relative water contents from 1.00 to 0.95 (Fig. 23), a relation which was found by means of a few iterations.

Especially between 12h30 and 13h00, there are considerable deviations between the measured and simulated results. These could be eliminated by introducing a curvilinear relation between conductance and relative water content, but neither the experimental data nor the physiological insight in the process warrants such a refinement.

The simulated stomatal conductance of sunlit leaves with a high relative water content is about half the maximum value of 0.0143 m s^{-1} (Fig. 23), because the stomatal opening is at high

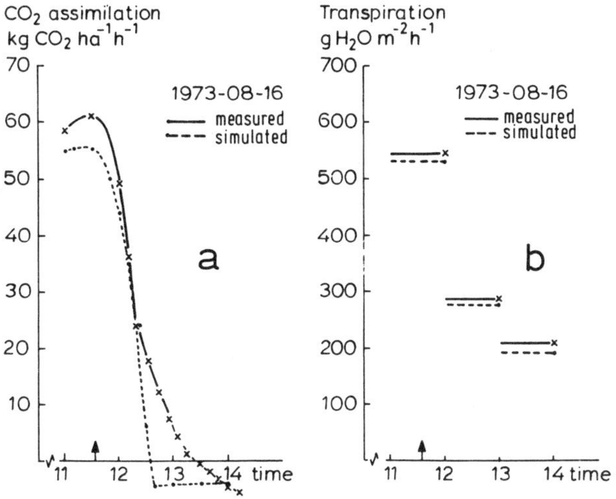

Fig. 22. CO_2 assimilation and transpiration of maize. Species: Zea mays cv. Caldera 535; density: 10 plants m^{-2}; measuring date: 1973–08–16; sowing date: 1973–05–01; location: Droevendaal, Wageningen; LAI: 5.3 m^2 m^{-2}; dry weight shoot: 12645 kg ha^{-1}; stage: 4.5 (Hanway, 1966); height: 280 cm (van Laar *et al.*, 1977).

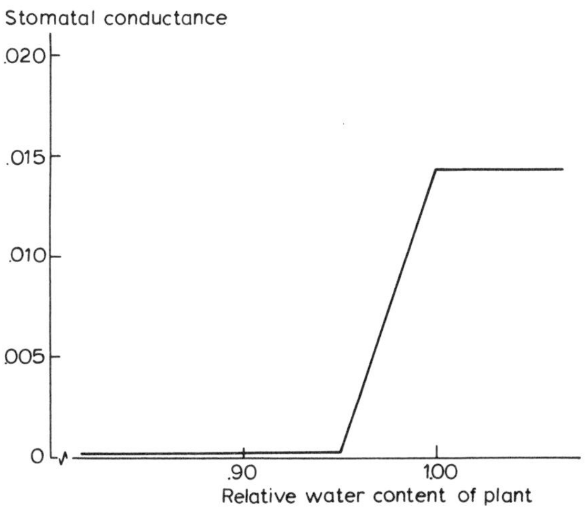

Fig. 23. The relation between relative water content and stomatal conductance in the absence of stomatal control by internal CO_2-concentration.

relative water contents controlled by the internal CO_2-concentration and not by the water content. The higher conductances manifest themselves only at lower external CO_2-concentrations.

These could also be seen in another experiment, where during a bright day and with plants in their tasselling stage, the CO_2 concentration in the enclosure was varied from 500 vppm to about 100 vppm, the results being presented in Figs 24a–c, as net assimilation and transpiration against time and net assimilation against CO_2 concentration. The measured and simulated net assimilation within the range of about 500–200 vppm CO_2 does not change. These results show indeed that the stomata are regulated in such a way that the internal CO_2-concentration remains constant within this trajectory, as illustrated in Fig. 24c for the simulated stomatal opening.

As has been explained before, a constant internal CO_2-concentration can only be maintained as long as the external CO_2-concentration is considerably higher than the setpoint of 120 vppm. At lower concentrations, a ratio of 0.4 for the external versus internal CO_2-concentration had to be introduced for the laboratory experiments (5.3) to explain better the observed stomatal regulation in the lower CO_2 range. When this ratio was used for the enclosure experiments, net assimilation started to decrease around an external CO_2-concentration of 300 vppm. By assuming a ratio of 0.6, this decline was shifted towards 200 vppm, but even then a considerable discrepancy remained with the experimental data. A further change of this ratio together with the assumption that the maximal stomatal conductance was larger than 0.0143 $m\,s^{-1}$, improved the situation somewhat, but led also to a too large discrepancy between simulated and experimental transpiration.

Further experimentation with the relevant parameters in the simulation program showed that the experimental results can only be understood if it is assumed that the mesophyll resistance decreases with decreasing internal CO_2-concentration, but this explanation is too speculative to be incorporated in the simulation program. Hence, the discrepancy between simulated and experimental CO_2-assimilation shows the lack of understanding. But the practical consequences are small because these low external CO_2-concentrations do not occur under field conditions.

Besides, experiments were done with young plants in their 4th leaf stage, a leaf area index of 10.7 being obtained by planting at a distance of about 4×4 cm. The simulated and measured dependence of net assimilation on CO_2-concentration is given in Fig. 24f. The

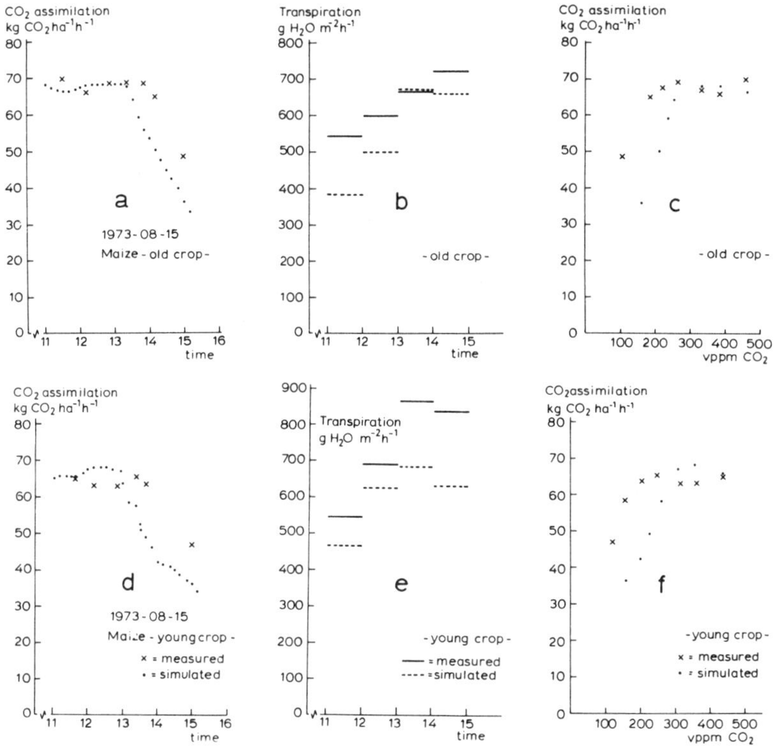

Fig. 24. CO_2 assimilation and transpiration of maize as a function of time and CO_2 assimilation as a function of CO_2 concentration of an old (o) and young (y) crop. Species: Zea mays cv. Caldera 535; density: 10 plants m^{-2}; measuring date: 1973–08–15; sowing date: 1973–05–01(o); 1973–07–11 (y); location: Droevendaal, Wageningen; LAI: 10.7 (y), 5.3 (o) $m^2 m^{-2}$; dry weight shoot: 3403 (y); 12645 (o) kg ha^{-1}; stage: 0.5–1.0 (y), 4.5 (o) (Hanway, 1966); height: 50 (y), 280 (o) cm.
N.B.: The seeds for the young crop were planted at a density of about 4×4 cm to ensure a closed crop surface at a young stage. Hence the LAI is extremely high (van Laar *et al.*, 1977).

simulation was done here with a maximum assimilation rate of 50 kg CO_2 ha^{-1} h^{-1}, the maximum rate being lower for these very young plants. Here again, the simulated decrease in assimilation occurs at a higher external CO_2-concentration than the measured decrease. Again it was not possible to achieve a proper match without speculating about a decreasing mesophyll resistance with decreasing internal CO_2-concentration.

On 22 August 1973, an elaborate experiment was done with maize planted at a normal density on 1 May (old plants) and with maize planted at a density of 4×4 cm on the 11 July (young plants). During the day, the enclosures were covered at four intervals to measure dark respiration, whereas the temperature was varied within the range of 10–30°C, as shown in Figs 25c and f. The measured and simulated results for both situations are presented in Figs 25a and d. The results show a good agreement of the dark respiration, within the normal temperature range, but at 30°C dark respiration is underestimated by simulation. The simulation could be improved in this particular case by assuming either a larger Q_{10} for maintenance in the 20–30°C range or a larger effect of temperature on the relative consumption rate of reserves. However, comparison of other simulated and measured results showed that sometimes respiration is underestimated and sometimes overestimated without any apparent reason.

Both simulations were done with a maximum assimilation rate of 70 kg CO_2 ha^{-1} h^{-1} which leads to underestimation for the old crop and overestimation for the young crop. These deviations could be easily eliminated by adapting this maximum rate, but this is a fruitless exercise because other experiments would then need other adjustments. It should be realized that for the densily planted young maize crop, the amount of roots were adapted. When this adaption was omitted, the plants developed water stress and the simulated results were governed by stomatal closure only.

Fig. 25. CO_2 assimilation and transpiration of an old and young maize crop. Species: Zea mays cv. Caldera 535; measuring date: 1973–08–22; location: Droevendaal, Wageningen. Young crop: sowing date: 1973–07–11; LAI = 15.5 m^2 m^{-2}; dry weight shoot: 5010 kg ha^{-1}; stage: 0.5–0.1 (Hanway, 1966); height: 75 cm. Old crop: sowing date: 1973–05–01; LAI: 5.7 m^2 m^{-2}; dry weight shoot: 16300 kg ha^{-1}; stage: 4.5 (Hanway, 1966); height: 280 cm; density: 10 plants m^{-2} (van Laar *et al.*, 1977).

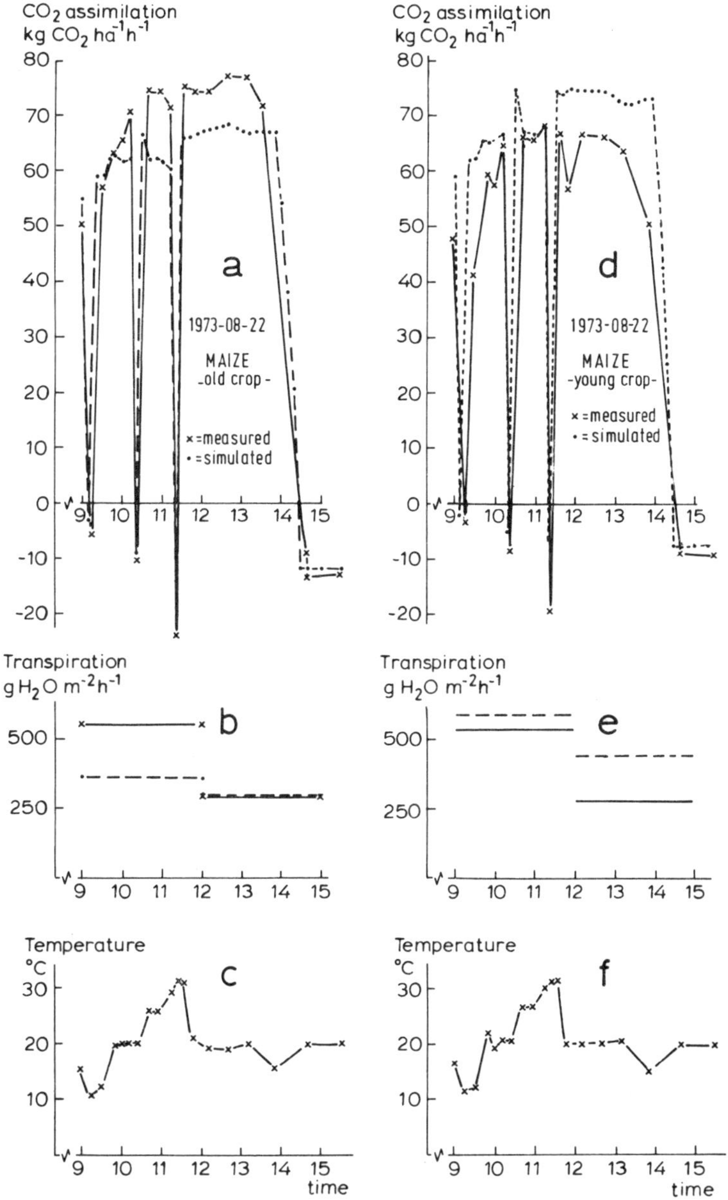
CO2 assimilation
kg CO2 ha-1 h-1
a
1973-08-22
MAIZE
-old crop-
× = measured
• = simulated
d
1973-08-22
MAIZE
-young crop-
× = measured
• = simulated
Transpiration
g H2O m-2 h-1
b
e
Temperature
°C
c
f
time

The simulated and actual daily course of transpiration are presented in Figs 25b and e as three-hour averages. This averaging was done because of the large time-lag in collecting the water from the cooler in combination with the large discontinuities in temperature and light regime. Because of the primitive method of measuring transpiration, a more satisfactory agreement could not be obtained. Therefore a much better method for measuring transpiration was adopted (see 2.2.2).

Detailed comparison of the two measured assimilation curves in

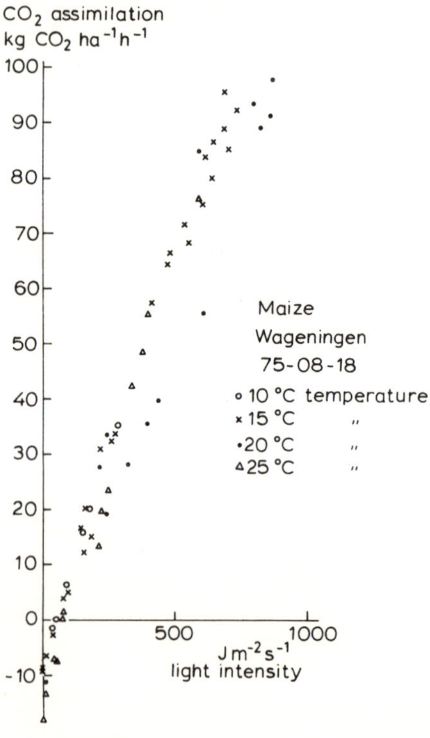

Fig. 26. CO_2 assimilation versus light intensity for field grown maize, measured in enclosures at four temperatures. Species: Zea mays cv. Caldera 535; density: 18 plants m^{-2}; measuring date: 1975–08–18; sowing date: 1975–04–20; location: Droevendaal, Wageningen; LAI: 7.76 m^2 m^{-2}; dry weight shoot: 18 687 kg ha^{-1}; stage: 5 (Hanway, 1966; van Laar *et al.* 1977).

Figs 25a and d with the temperature curves reveals that response to temperature is different. Old plants hardly react to a change in temperature, whereas the assimilation of young plants goes up and down with this variable. Other measurements in Fig. 26 show that the temperature effect on the assimilation of old plants is small. Under these conditions, the results of the simulation of temperature experiments are in fair agreement with the measurements, if it is assumed that maximum leaf assimilation is practically independent of temperature above about 13°C, as was done in both cases.

This lack of temperature response was not observed under controlled conditions, either because no measurements were done with mature leaves of plants in their tasselling stage or because constant

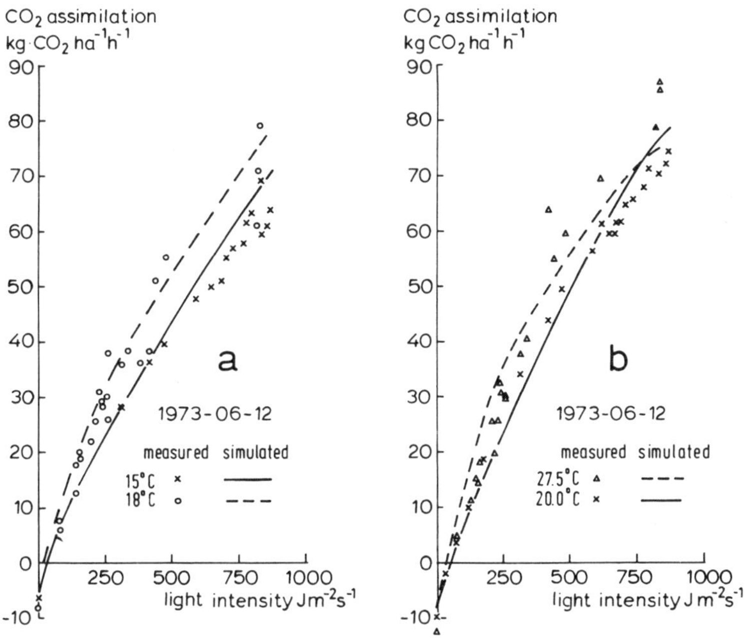

Fig. 27. CO_2 assimilation at four temperatures as a function of light. Species: Zea mays cv. Caldera 535; density 3×3 cm (a), 4×4 cm (b); measuring date; 1973–06–12; sowing date: 1973–05–01; location: Droevendaal, Wageningen; LAI: 17.5 $m^2 m^{-2}$ (a), 14.7 $m^2 m^{-2}$ (b); dry weight shoot: 3708 kg ha^{-1} (a), 3297 kg ha^{-1} (b); stage: 0.5–1.0 (Hanway, 1966; van Laar *et al.*, 1977).

environmental conditions do not induce this temperature adaptation. The phenomenon has not been studied in sufficient detail to discuss the physiological and biochemical background, mainly because it escaped the attention of crop scientists. However, already in the 1930s, extension officers in the Netherlands came to the conclusion that maize performs exceedingly well in autumn at relatively low temperatures.

As has been said, the response of the young plants to temperature changes is more pronounced. This has been analysed in more detail in a comparative experiment. The results are presented as a function of light intensity in Figs 27a and b. Here, the crop assimilation appears to vary more than 10% with the induced temperature variation of 5 to 7°C on either side of 20°C. Since the initial efficiency of light use is considered to be independent of temperature within the range concerned, these changes can be only simulated by assuming that the effect of temperature on maximum assimilation of individual leaves is about twice as large.

Some trial runs showed that the maximum assimilation has to be estimated at 37.5, 50 and 70 kg CO_2 ha^{-1} h^{-1} for temperatures of 15, 20 and 30°C, respectively. The curves in Fig. 27 are the simulated results. These agree with the measured results, except that at high light intensities at 27.5°C, the simulated results are too low. This difference is due to an increasing water shortage in the simulation program. This temperature effect is of the same magnitude as the temperature effect based on measurements of mature leaves of plants in about their sixth leaf stage and grown under controlled conditions, as given in Fig. 14.

8.3 Field experiments

The changes, adopted in the previous section – with a maximum rate of assimilation of 70 kg CO_2 ha^{1} h^{-1} at temperatures above 13°C –, were also incorporated in BACROS and subsequent field evaluations were done without any further changes or adaptations, except for the forcing functions.

Field experiments in Wageningen, Flevoland (the Netherlands), Davis (California) and Ames (Iowa) are used for evaluation. The experiments in Wageningen and in Flevoland were done by L. Sibma from the Centre for Agrobiological Research, in Davis (Cal.) at the Department of Agriculture under the guidance of W. A. Williams and C. T. de Wit and the result of the experiment in Ames (Iowa) was kindly made available by R. M. Shibles. Some charac-

Table 4. Mean weather data and crop growth rates of maize in its grand period of growth at four locations.

Average for July					
Location	latitude	radiation $Jm^{-2}\ day^{-1}$	max temp. °C	min temp. °C	crop growth rate $kg\ ha^{-1}\ day^{-1}$
Flevoland					
1971	52	17.82×10	22.76	11.56	250
1972	52	15.24×10^6	20.82	13.26	210
Wageningen					
1976	52	19.14×10^6	25.21	12.79	210
Ames					
1963	42	23.2×10^6	29.68	17.55	300
Davis					
1968	39	30.8×10^6	34.35	12.47	350

teristic weather data for the three locations are given in Table 4, together with the average crop growth rate during the grand period of growth. Detailed information on leaf area and dry matter weights are given in Table 5. The experiments in the Netherlands were done in 1971, 1972 and 1976, in Davis in 1968 and in Ames in 1963. Planting densities covered a range from 2.5 to 40 plants m^{-2} and the crops were periodically harvested at 6–8 dates.

The experimental and simulated results in kg dry matter ha^{-1} are presented in Figs 28–31 as a function of time. The observational points are the mean yields. The standard deviation of these mean yields is about 10%. The continuous curves represent the simulated results. These are obtained by introducing the actual weather data as forcing functions. The courses of the leaf area index and the chemical composition with time are derived from the experimental results and are also used as forcing functions, because our aim is to evaluate the simulation of the dry matter accumulation process only. It would be possible to relate leaf area growth with crop growth and development and simulate in this way the whole growth process, but this was not our intention as was discussed in 2.1.

The agreement between simulated and measured results is in general good, except for the low density in Davis. The simulated dry matter yield is here about half the measured yield. It is quite incredible that the actual assimilation is so much higher than the simulated assimilation with a maximum rate of 70 kg CO_2 $ha^{-1}\ h^{-1}$. The measured leaf area index at this low density must have been too low. This assumption is supported by the observation that at the low

Table 5. Data of dry shoot weight and leaf area index on the harvest days at four locations.

Davis 1968

harvest day	dry matter kg ha^{-1}					leaf area index m^2 m^{-2}					
	2.5	5	10	20	40	2.5	5	10	20	40	plants m^{-2}
13 June	25	52	103	175	345	0.05	0.11	0.21	0.36	0.76	
27 June	465	722	1221	2201	2834	0.59	1.10	2.09	3.90	5.92	
11 July	3572	4690	6244	6781	7886	1.50	2.90	4.99	8.32	12.20	
25 July	8107	9800	11177	13572	14874	1.65	3.39	6.05	9.70	13.76	
8 Aug.	11650	14480	14958	18366	17704	1.69	3.18	5.53	9.17	9.90	
12 Oct.	12752	14116	16779	18487	20619	—	—	—	—	—	

Ames 1963

harvest day	dry matter kg ha^{-1}					leaf area index m^2 m^{-2}					
	2.42	3.16	4.32	6.18	9.69	2.42	3.16	4.32	6.18	9.69	plants m^{-2}
12 June	108	141	191	248	391	0.15	0.20	0.28	0.36	0.57	
19 June	305	428	591	821	1185	0.34	0.49	0.69	0.98	1.45	
26 June	866	1118	1574	2210	2906	0.78	1.04	1.52	2.23	3.03	
13 July	1829	2446	2943	3729	4335	1.32	1.84	2.34	3.15	3.88	
10 July	3004	3342	4268	5104	6023	1.72	2.02	2.80	3.63	4.61	
22 July	5271	6874	8130	8976	10540	1.91	2.80	3.72	4.61	6.05	
8 Aug.	10051	11143	13023	14483	17413	1.97	2.92	3.52	4.61	6.40	
29 Aug.	11888	13656	16622	17687	19254	1.46	2.22	3.04	3.81	4.17	
17 Sept.	13258	16325	17926	20226	22813	1.19	1.82	2.38	2.87	2.36	

Table 5. (continued)

Flevoland 1971[1] harvest day	d.m. kg ha^{-1}	LAI $m^2 m^{-2}$	**Flevoland 1972**[1] harvest day	d.m. kg ha^{-1}	LAI $m^2 m^{-2}$
1 June	10	0.180	22 June	48	0.086
15 June	430	0.800	5 July	217	0.409
29 June	1360	1.800	19 July	1082	2.034
13 July	5490	5.184	26 July	3082	3.943
27 July	9768	5.656	7 Aug.	5681	3.876
10 Aug.	10698	5.103	16 Aug.	6647	3.429
24 Aug.	14446	4.638	28 Aug.	10013	3.444
13 Sept.	17095	3.957	11 Sept.	12012	3.2

Wageningen 1976[1] **– Capella –** harvest day	d.m. kg ha^{-1}	LAI $m^2 m^{-2}$	**Wageningen 1976**[1] **– P 3853 –** d.m. kg ha^{-1}	LAI m m^{-2}
24 May	12.7	0.029	7	0.019
17 June	201	0.39	130	0.24
13 July	5420	3.71	4390	4.11
8 Aug.	12000	3.84	10146	4.18
7 Sept.	17870	3.39	20560	3.84
21 Sept.	21630	3.23	20910	4.55

[1] Plant density: 10 plants m^{-2}

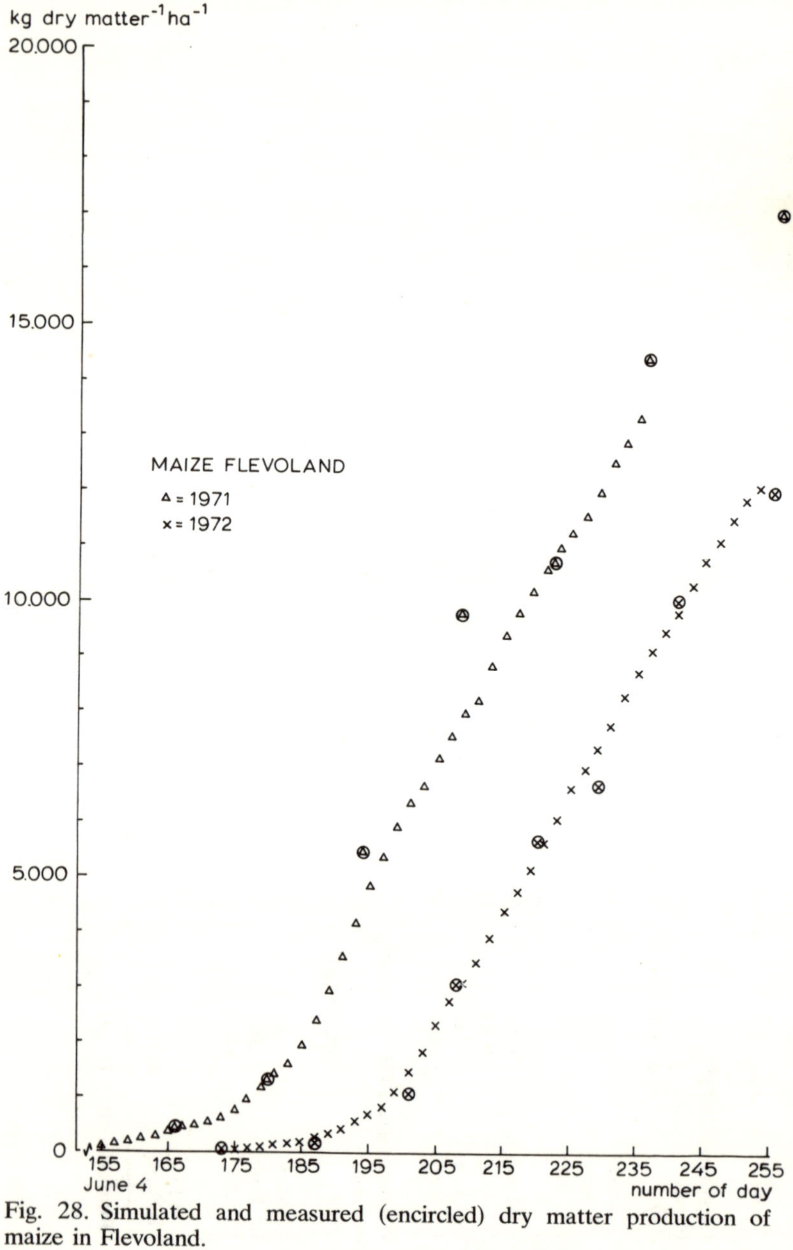

Fig. 28. Simulated and measured (encircled) dry matter production of maize in Flevoland.

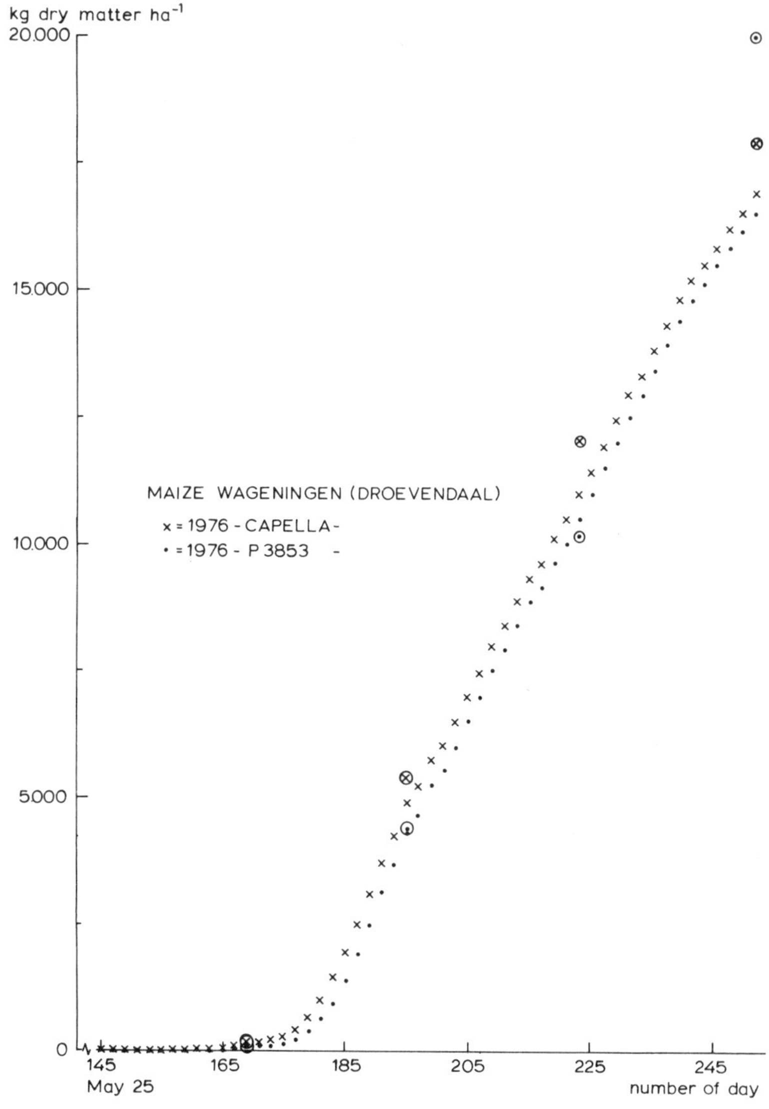

Fig. 29. Simulated and measured (encircled) dry matter production of maize in Wageningen.

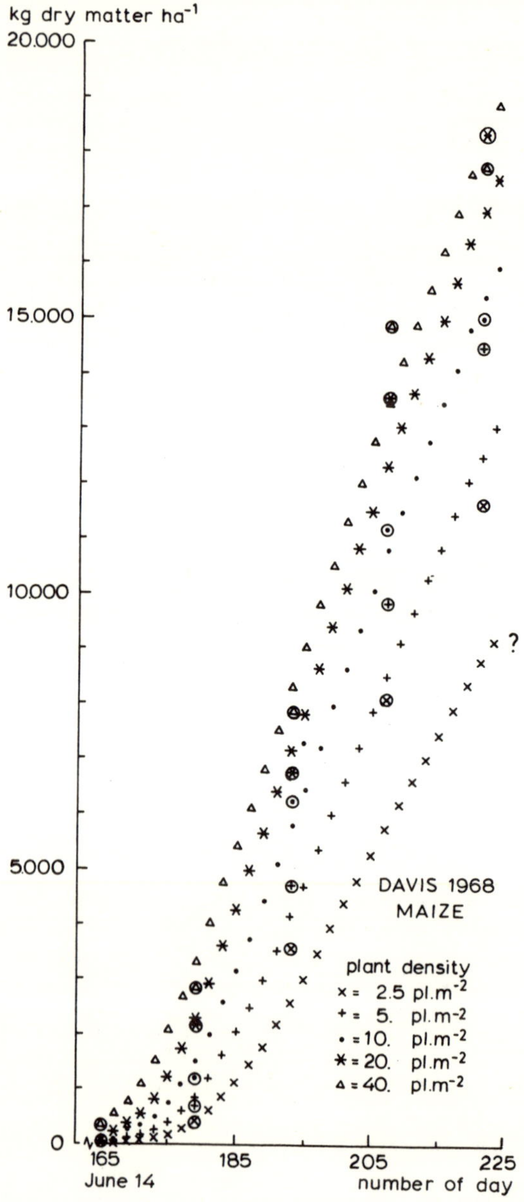

Fig. 30. Simulated and measured (encircled) dry matter production of maize with different plant densities in Davis.

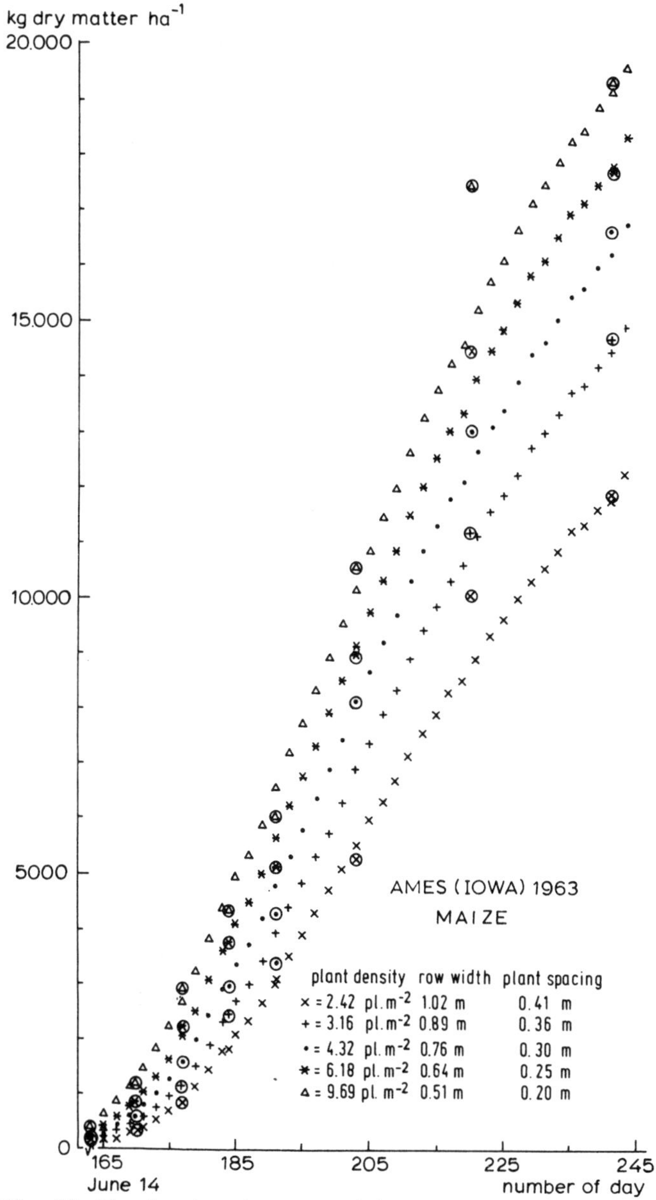

Fig. 31. Simulated and measured (encircled) dry matter production of maize with different plant densities in Ames.

density in Davis the leaf area ratio (area/weight crop) is about half that at the high density. But this observation is not confirmed in Ames. Deviations may also be attributed to the estimation of the exposed leaf area index, since a part of the leaf area of young plants is not exposed to the light but enclosed in other leaves and leaf sheaths, as said before.

The differences in the experimental conditions with respect to latitude, radiation and temperature and with respect to planting density cover a large range, which appears also from the crop growth rate which ranged from about 210 to 350 kg ha^{-1} day^{-1}. Therefore, the simulation program BACROS, as presented in Appendix A, is used with some confidence to simulate potential dry matter accumulation rates of maize in feasible growth situations. Consequently, further development is not so much directed towards the improvement of the present program, but towards the simulation of morphogenesis, especially in the early phase of growth and after tasselling.

8.4 Root growth

Several attempts were made to determine the amount of dry matter accumulated in the roots in the field situation, but the problem of achieving sufficient accuracy without investing a prohibitive amount of work has not been solved. The conclusion from experiments is that, stubble excluded, about 1500 kg ha^{-1} roots accumulate and by adjusting the ratio between root conductance and root biomass, the program has been rigged in such a way that these amounts of roots are simulated in the Wageningen situation.

It has been shown by a mathematical analysis in Chapter 6 that the simulated root-shoot ratio increases about proportionally with the square root of the evaporative demand and that the time constant of adaptation of the root-shoot ratio is also realistic. Both phenomena are illustrated here by some simulation runs with 1972 Flevoland data.

In Fig. 32 the simulated relation between shoot and root weight is presented for the normal situation and for situations where either half of the roots is removed, or the root weight is doubled. The original shoot-root ratio is gradually restored as the weight increases. However, full restoration does not occur because changes in root weight also affect the ratio between young and suberized roots and hence the average root conductance.

Fig. 33 presents the simulated shoot versus root weight under

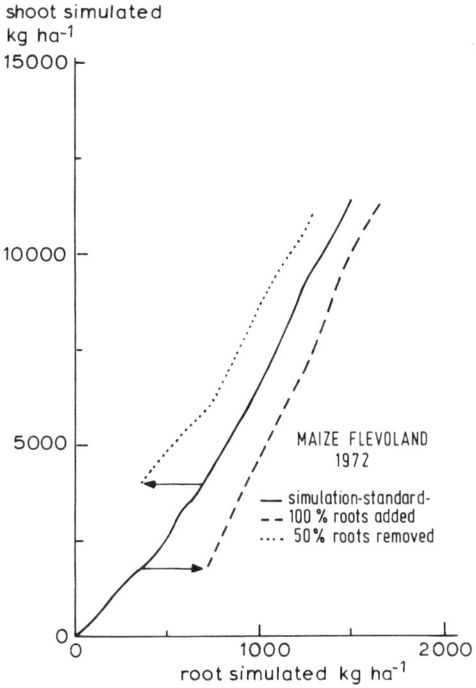

Fig. 32. Simulated shoot-root ratio for maize; the root weight is doubled at 350 kg ha^{-1} and halved at 700 kg ha^{-1}. —— = simulation standard; - - - = 100% roots added; · · · = 50% roots removed.

different evaporative demands which were created in the simulation program by varying the humidity of the air. As suggested in 6.5 the root-shoot ratio indeed increases about proportionally with the square root of the transpiration. This behaviour is assumed to be more or less realistic, but conclusive data are lacking.

Fig. 34 presents the simulated shoot versus root growth in Davis, Ames and Wageningen for a normal plant density. However probable these differences seem, there are again no data to show that they actually occur.

8.5 Transpiration

Under the conditions of Fig. 33 the potential evapotranspiration

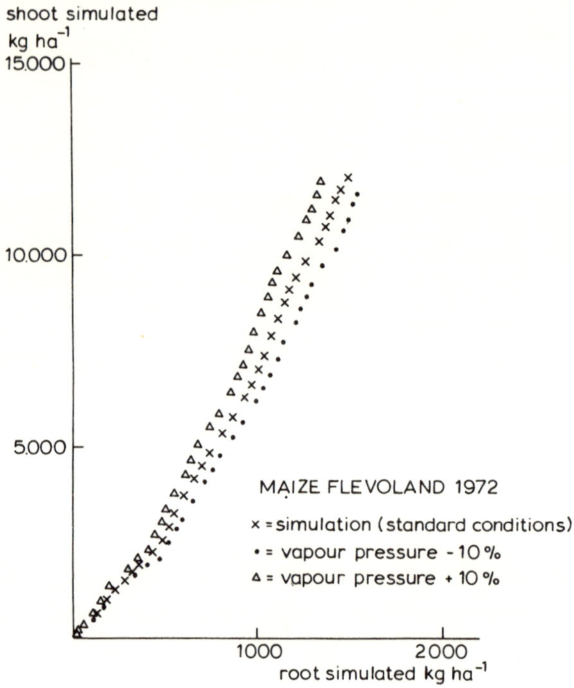

Fig. 33. The influence of a difference in vapour pressure on the shoot-root ratio. The potential evaporation at a vapour pressure of +10, 0 and −10 percent was 4.2, 4.7 and 5.1 mm day^{-1} and the transpiration (exclusive evaporation from the soil) was 1.18, 1.44 and 1.71 mm day^{-1}, respectively.

was 5 mm day^{-1}, whereas the transpiration was only 1.4 mm day^{-1}. Even when we take into account that this figure does not include the evaporation from the soil, this transpiration is surprisingly small. This is a direct consequence of the assumption that the stomatal conductivity is governed by net assimilation in such a way that the internal CO_2-concentration is maintained at 120 vppm. When this assumption is relaxed so that all stomata are only closed during night and open during the day at 0.008 m s^{-1}, the daily transpiration is 5.4 mm day^{-1}, a seemingly more normal value.

However, the simulated transpiration in PHOTON then becomes too large, and the simulated values of the temperature and humidity differences above and inside the crop deviate from the observations

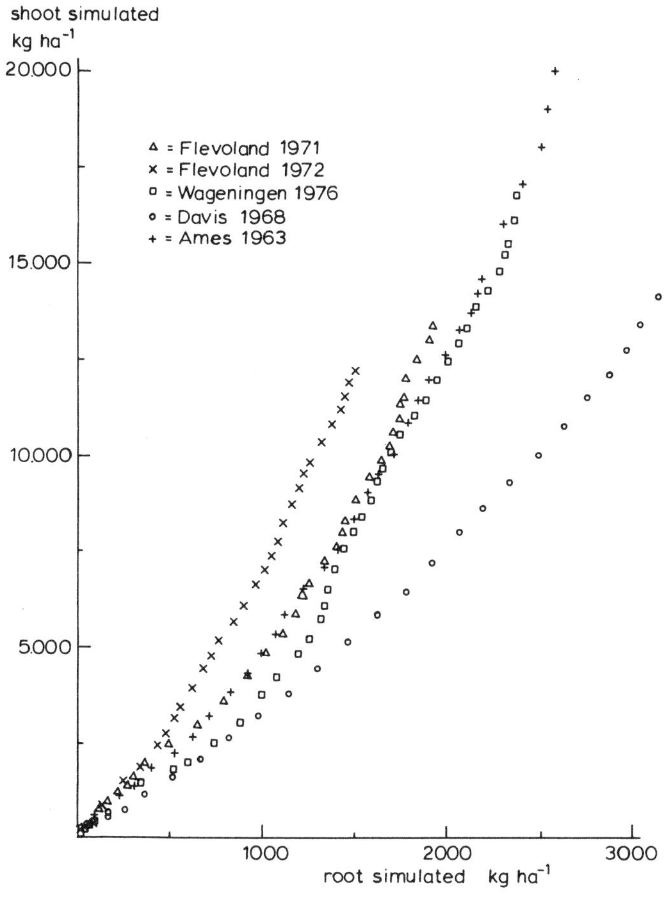

Fig. 34. Simulated shoot-root ratios at different locations.

as can be seen from the results presented in Fig. 35. Moreover the assumption of open stomata is completely contradicted by observations with the diffusion porometer, as illustrated in Fig. 36.

From all this evidence, it is concluded that this low transpiration rate of maize is realistic, however surprising this may be. Together with a crop growth rate of about 200 kg ha^{-1} day^{-1}, this leads to a transpiration coefficient smaller than 100 kg water per kg dry mat-

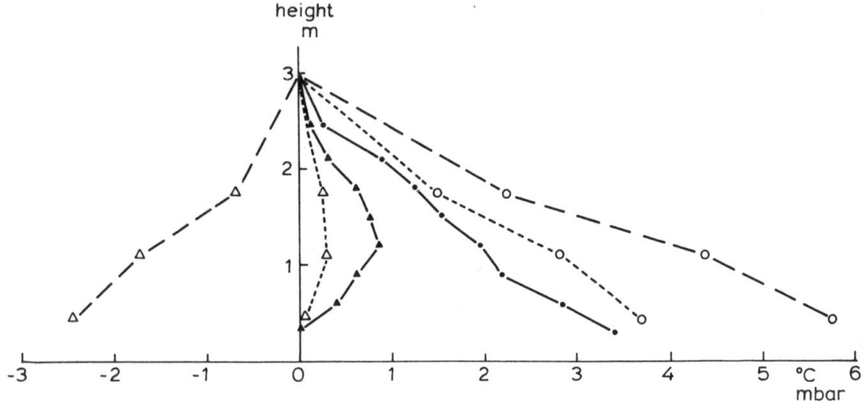

Fig. 35. Simulated and measured profiles of air temperature and humidity in a maize crop on 14 August 1973. The differences with respect to the value at 3 m height are plotted against height above the soil surface.

	temperature	humidity
Measured	▲——▲	●——●
Simulated with regulated stomata	△----△	○----○
Simulated with stomata fully open	△——△	○——○

ter. Obviously, nature has developed a mechanism which allows the combination of high growth rates with small transpiration rates. This mechanism is of extreme importance for agriculture in (semi)-arid regions.

But how common is this phenomenon? Brown (1964) measured by the flux method an evapo-transpiration of somewhat below 3 mm day^{-1} and an evaporation of somewhat below 1 mm day^{-1} for maize on a clear day. One may doubt the accuracy of the measuring method, but these low figures were also simulated by Goudriaan (1977) assuming internal CO_2-regulation. Experiments of Musgrave and his students, Cornell University, (pers. com.) show consistently that net assimilation of the leaves of field grown maize is over a large range independent of CO_2 concentration, which is also a strong indication of regulation of the internal CO_2-concentration. However, experiments with maize grown under controlled conditions, reported in literature cover the whole range from the absence of any effect to a proportional effect of CO_2 concentrations. Thus

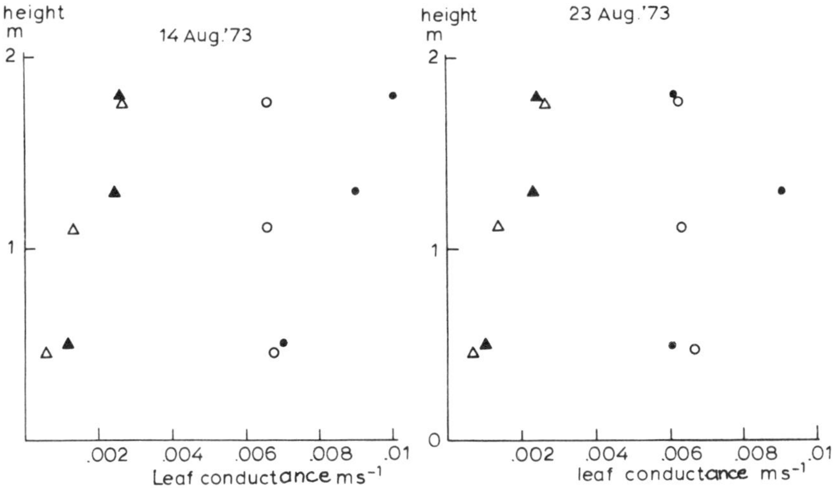

Fig. 36. Measured (▲ and ●) and simulated (△ and ○) leaf conductances for water vapour, on 14 August 1973 (a) and 23 August 1973 (b) for shaded (△ and ▲) and sunlit leaves (○ and ●). For measurements and further details see Stigter & Lammers (1974).

these results point to the whole range from regulation to non regulation of the internal CO_2-concentration within the same species, in dependence of variety, growing conditions or both.

Our own experiments on sunflower under controlled conditions show that the stomatal opening is independent of net assimilation (Fig. 15a), but experiments in enclosures show again a close relation between transpiration and net assimilation, and thus regulation of internal CO_2-concentration. The analyses of enclosure experiments with wheat indicate an intermediate position for this crop. As has been said in 2.2.2, the instantaneous measurement of transpiration has been perfected considerably, and it is therefore expected that more pertinent information will become available in due course.

The simulation program includes the assumption that the internal CO_2-concentration is regulated. The other extreme of open stomata during the day and possibly closed stomata during the night is easily programmed. The programming might be more difficult for intermediate situations. In the most general case, it has to be assumed that stomatal opening and net assimilation both depend on internal

CO_2-concentration. This assumption leads to a program section with the internal CO_2-concentration as a state variable (integral). If the resulting small time constant appears too troublesome, this state variable has to be eliminated by introducing another implicit loop, as in 4.4.

References

Alberda, Th., *et al.*, 1977. Crop photosynthesis: methods and compilation of data obtained with a mobile field equipment. Versl. Landbouwk. Onderz. (Agr. Res. Rep.) 865, Pudoc, Wageningen, 46 pp.

Angström, A., 1924. Report to the international commission for solar research on actinometric investigations of solar and atmospheric radiation. Quart. J. Roy. Met. Soc. 50: 121–126.

Barrs, H. D., 1970. Controlled environment studies of the effects of variable atmospheric water stress on photosynthesis, transpiration and water status of Zea mays L. and other species. UNESCO Symposium on Plant Response to Climatic Factors, Uppsala.

Björkman, O., 1966. The effect of oxygen concentration on photosynthesis in higher plants. Phys. Plant. 19: 618–633.

Björkman, O. & J. Ehleringer, 1975. Comparison of the quantum yields for CO_2 uptake in C_3 and C_4 plants. Carnegie Institution Year Book 74: 760–761.

Brouwer, R., 1963. Some aspects of the equilibrium between overground and underground plant parts. Jaarboek Instituut voor Biologisch en Scheikundig onderzoek (IBS) 31–39.

Brouwer, R., 1965. Water movement across the root. Symp. Soc. Exp. Biol. 19: 131–149.

Brouwer, R. & C. T. de Wit, 1968. A simulation model of plant growth with special attention to root growth and its consequences. Proc. 15th Easter School Agric. Sci., Univ. of Nottingham, 224–242.

Brown, K. W., 1964. Vertical fluxes within the vegetative canopy of a corn field. U.S. Dept. of Agriculture and Cornell University, Ithaca, N.Y. Interim rep. 64–1.

Brunt, D., 1932. Notes on radiation in the atmosphere. I. Quart. J. Roy. Met. Soc. 58: 389–420.

Challa, H., 1976. An analysis of the diurnal course of growth, carbon dioxide exchange and carbohydrate reserve content of cucumber. Versl. Landbouwk. Onderz. (Agr. Res. Rep.) 861, Pudoc, Wageningen, 88 pp.

Charles-Edwards, D. A. & J. L. Ludwig, 1974. A model for leaf photosynthesis by C_3 plant species. Ann. Bot. 38: 921–930.

Dayan, E. & A. Dovrat, 1977. Measured and simulated herbage production of Rhodes grass. Internal Report of the Hebrew University of Jerusalem, Department of Field and Vegetable Crops, Faculty of Agriculture, Rehovot, Israel.

Dijkshoorn, W., 1971. Partition of ionic constituents between organs. In: Recent Advances in Plant Nutrition. R. M. Samish Ed., Gordon and Breach Science Publishers Inc., New York, 447–476.
Ehleringer, J. & O. Björkman, 1976. Carbon dioxide and temperature dependence of the quantum yield for CO_2 uptake in C_3 and C_4 plants. Carnegie Institution Year Book 75: 418–421.
Goudriaan, J., 1977. Crop micrometeorology: a simulation study. Simulation Monographs. Pudoc, Wageningen, 249 pp.
Goudriaan, J. & H. H. van Laar, 1978. Measurement of some relations between leaf resistance in maize, beans, lalang-grass and sunflower (In press, Photosynthetica 12.)
Hanway, J. J., 1966: How a corn plant develops. Spec. Rep. 48, Iowa State University.
Keulen, H. van, 1975. Simulation of water use and herbage growth in arid regions. Simulation Monographs. Pudoc, Wageningen, 184 pp.
Keulen, H. van, 1976. A calculation method for potential rice production. Contr. Centr. Res. Inst. Agric. Bogor, No. 21: 26 pp.
Keulen, H. van & W. Louwerse, 1975. Simulation models for plant production. Proc. W.M.O. Symposium 'On agrometeorology of the wheat crop'. Braunschweig, Deutscher Wetterdienst, Offenbach a.M.
Keulen, H. van, W. Louwerse, L. Sibma & Th. Alberda, 1975. Crop simulation and experimental evaluation: a case study. In: Photosynthesis and productivity in different environments. Ed. J. P. Cooper. Cambridge University Press, 623–643.
Kleinendorst, A. & R. Brouwer, 1972. The effect of local cooling on growth and water content of plants. Neth. J. agric. Sci. 20: 203–217.
Kuiper, P. J. C., 1964. Water uptake of higher plants as affected by root temperature. Meded. LandbHogesch. Wageningen, 64–4.
Kuiper, P. J. C., 1972. Water transport across membranes. Ann. Rev. Pl. Physiol. 23: 157–172.
Laar, H. H. van & F. W. T. Penning de Vries, 1972. CO_2-assimilation light response curves of leaves; some experimental data. Versl. IBS, 62, Wageningen.
Laar, H. H. van, D. Kremer & C. T. de Wit, 1977. Maize. In: Crop photosynthesis: methods and compilation of data obtained with a mobile field equipment. Ed. Th. Alberda. Versl. Landbouwk. Onderz. (Agr. Res. Rep.) 865, Pudoc, Wageningen, 12–21.
Lambert, J. R. & F. W. T. Penning de Vries, 1973. Dynamics of water in the soil-plant-atmosphere system: A model named TROIKA. In: Ecological studies. Analysis and Synthesis, Vol. 4. A. Hadas *et al.* (Eds.), Springer Verlag, Berlin, 257–273.
Leafe, E. L., 1972. Micro-environment, carbon dioxide exchange and growth in grass swards. In: Crop processes in controlled environments. Eds. A. R. Rees, K. E. Cockshull, D. W. Hand & R. J. Hurd. Academic Press, London, 157–174.
Lof, H., 1976. Water use efficiency and competition between arid zone

annuals, especially the grasses Phalaris minor and Hordeum murinum. Versl. Landbouwk. Onderz. (Agr. Res. Rep.) 853, Pudoc, Wageningen, 109 pp.

Louwerse, W. & J. W. Eikhoudt, 1975. A mobile laboratory for measuring photosynthesis, respiration and transpiration of field crops. Photosynthetica 9, 31–34.

Monteith, J. L., 1973: Principles of environmental physics. Edward Arnold, London, 241 pp.

Neales, T. F. & L. D. Incoll, 1968. The control of leaf photosynthesis rate by the level of assimilate concentration in the leaf. A review of the hypothesis. Bot. Rev., Vol. 34, 2.

Pearman, G. I., H. L. Weaver & C. B. Tanner, 1972. Boundary layer heat transfer coefficients. Agric. Met. 10: 83–93.

Penman, H. L., 1948. Natural evaporation from open water, bare soil and grass. Proc. Roy. Soc. A., 193: 120–146.

Penning de Vries, F. W. T., 1972. Respiration and growth. In: Crop processes in controlled environments. Eds. A. R. Rees, K. E. Cockshull, D. W. Hand & R. J. Hurd. Academic Press, London, 327–347.

Penning de Vries, F. W. T., 1975a. Use of assimilates in higher plants. In: Photosynthesis and productivity in different environments. Ed. J. Cooper, Cambridge Univ. Press, 459–480.

Penning de Vries, F. W. T., 1975b. The cost of maintenance processes in plant cells. Ann. Bot. 39: 77–92.

Penning de Vries, F. W. T., 1977c. Evaluation of simulation models in agriculture and biology: conclusions of a workshop. Agric. Systems, 2: 99–107.

Penning de Vries, F. W. T., A. H. M. Brunsting & H. H. van Laar, 1974. Products, requirements and efficiency of biosynthesis: a quantitative approach. J. theor. Biol. 45: 339–377.

Penning de Vries, F. W. T. & H. H. van Laar, 1977a. Substrate utilization in germinating seeds. In: Environmental effects on crop physiology. Eds. J. J. Landsberg & C. V. Cutting. Academic Press, London, 217–228.

Penning de Vries, F. W. T., J. M. Witlage & D. Kremer, 1979. Rates of respiration and of increase in structural dry matter in young wheat, ryegrass and maize plants in relation to temperature, to water stress and to their sugar content (in prep.).

Raschke, K., 1975. Stomatal action. Ann. Rev. Plant Physiol. 26: 309–340.

Sinclair, T. R. & C. T. de Wit, 1976. Analysis of the carbon and nitrogen limitations to soybean yield. Agronomy Journal 68: 319–324.

Sinclair, T. R., J. Goudriaan & C. T. de Wit, 1977. Mesophyll resistance and CO_2 compensation concentration in leaf photosynthesis models. Photosynthetica (11) 1, 56–65.

Stigter, C. J., J. Goudriaan, F. A. Bottemanne, J. Birnie, J. G. Lengkeek & L. Sibma, 1977. Experimental evaluation of a crop climate simulation model for Indian corn (Zea mays L.). Submitted to Agric. Met.

Stigter, C. J. & B. Lammers, 1974. Leaf diffusion resistance to water

vapour and its direct measurement. III. Results regarding the improved diffusion porometer in growth rooms and fields of Indian corn (Zea mays). Meded. LandbHogesch. Wageningen 74–21: 1–76.
Tadmor, N. H., M. Evenari & L. Shanan, 1972. Primary production of pasture plants as function of water use. In: Eco-physiological foundation of ecosystems productivity in arid zone. Int. Symposium, USSR Academy of Sciences. Publishing house 'NAUKA', Leningrad Branch, Leningrad, 151–157.
Tooming, H., 1967. Mathematical model of plant photosynthesis. Photosynthetica 1, 233–240.
Wit, C. T. de, 1958. Transpiration and crop yields. Versl. Landbouwk. Onderz. (Agr. Res. Rep.) 64.6, Pudoc, Wageningen, 88 pp.
Wit, C. T. de, 1965. Photosynthesis of leaf canopies. Versl. Landbouwk. Onderz. (Agr. Res. Rep.) 663, Pudoc, Wageningen, 57 pp.
Wit, C. T. de, 1970. Dynamic concepts in biology. In: Prediction and measurement of photosynthetic productivity. Proc. IBP/PP technical meeting. Trebon, 1969. Ed. I. Setlik. Pudoc, Wageningen, 17–23.
Wit, C. T. de & Th. Alberda, 1961. Transpiration coefficient and transpiration rate of three grain species in growth chambers. Jaarboek IBS 1961, 73–81.
Wit, C. T. de, R. Brouwer & F. W. T. Penning de Vries, 1970. The simulation of photosynthetic systems. In: Prediction and measurement of photosynthetic productivity. Proc. IBP/PP technical meeting. Trebon, 1969. Ed. I. Setlik. Pudoc, Wageningen, 47–50.
Wit, C. T. de, & H. van Keulen, 1972. Simulation of transport processes in soils. Simulation Monographs. Pudoc, Wageningen, 109 pp.
Wit, C. T. de & J. Goudriaan, 1974. Simulation of ecological processes. Simulation Monographs. Pudoc, Wageningen, 167 pp.

Appendix A – BACROS: Basic crop simulator

```
TITLE                         BASIC CROP SIMULATOR
*                             ===== ==== =========

/      DIMENSION Z(9,10), S(9,10)
*   Z: DISTRIBUTION OF THE LEAVES WITH RESPECT TO THE INCOMING SUNRAYS

*************************         SECTION  1      *****************************

*          MACRO DEFINITIONS

***** 1. AVERAGE TEMPERATURE OF CANOPY, SENSIBLE AND LATENT HEAT LOSS

MACRO TEHL,TSHL,AVTCP,NCRL=TRPH(VIS,NIR,LWR,AREA)
      ABSRAD=VIS+NIR+LWR
      EVA    =AMIN1(EFF*VIS/AMAX,46.)
*     PREVENTS UNDERFLOW
      NCRIL  =(AMAX+DPL)*(1.-EXP(-EVA) )-DPL
      SRESL  =(68.4*(ECO2C-RCO2I)-RA*1.32*NCRIL)/AMAX1(0.001,NCRIL)/1.66
      IF  (SRESL.GT.SRW.OR.SRESL.LT.0.) GO TO 700
      SRESL  =SRW
      NCRIL  =68.4*(ECO2C-RCO2I)/(SRW*1.66+RA*1.32)
  700 SRES   =AMIN1(RESCW,SRESL)
      ENP    =0.3*NCRIL
      EHL    =(SLOPE*(ABSRAD-ENP)+DRYP)/(PSCH*(RA*0.93+SRES)/RA+SLOPE)
      SHL    =ABSRAD-EHL-ENP
      TL     =TA+SHL*RR
      TEHL   =TEHL  +AREA*EHL
      TSHL   =TSHL  +AREA*SHL
      AVTCP  =AVTCP+AREA*TL
      NCRL   =NCRL  +AREA*NCRIL
ENDMAC
PARAM ECO2C=330.
*     EXTERNAL CO2-CONCENTRATION (VPPM)
PARAM RESCW=2000.
*     CUTICULAR RESISTANCE TO WATER, IN S/M
CONST PSCH =0.67
*     PSYCHROMETRIC CONSTANT IN MBAR PER KELVIN

***** 2. DAILY TOTALS

MACRO DTOT =DLYTOT(DTOTI,RATE)
      DTOT1=INTGRL(DTOTI,RATE)
      DTOT =DTOT1-ZHOLD(IMPULS(DELT,86400.)*KEEP,DTOT1)
*     THE ACCUMULATOR IS EMPTIED AFTER MIDNIGHT,
*     SO CONTENTS ARE AVAILABLE FOR PRINTING
ENDMAC

***** 3. FRACTIONS OF PLANT CONSTITUENTS IN INCREMENT

MACRO FIC=INCREM(FT)
      FIC=(AFGEN(FT,DAY+1.)*WSN-AFGEN(FT,DAY)*WSD)/(WSN-WSD)
ENDMAC

***** 4. INTERPOLATION OF TEMPERATURE ALONG SINE PROFILE

MACRO VAL    =WAVE(DAY,HOUR,SRTB,I4TB,RISE)
```

```
      TIM    =INSW(HOUR-14.,HOUR+10.,HOUR-14.)
*     MAXT   =AFGEN(I4TB,DAY+(HOUR-14.0)/24.)
*     MINT   =AFGEN(SRTB,DAY+(HOUR-RISE)/24.)
      MAXT   =AFGEN(I4TB,DAY+HOUR/24.)
      MINT   =AFGEN(SRTB,DAY+HOUR/24.)
      VALAV =0.5*(MAXT+MINT)
      VALAMP=0.5*(MAXT-MINT)
      VALSR =VALAV-VALAMP*COS(PI*(HOUR-RISE)/(14.-RISE) )
*     TEMP. DURING RISING OF SUN
      VALSS =VALAV+VALAMP*COS(PI*TIM/(10.+RISE) )
*     TEMP. DURING SETTING OF SUN
      VAL    =INSW(AND(HOUR-RISE,14.-HOUR)-0.5,VALSS,VALSR)
ENDMAC


************************          SECTION  2       ****************************

*        INITIALIZATION
INITIAL
FIXED I,J,K,L,N,IL,IS,ISUN,MAX,SN
STORAGE F(9), DAV(9)
* DAV: AVERAGE PROJECTION OF THE LEAVES IN THE DIFFERENT DIRECTIONS
TABLE F(1-9)=.015,.045,.074,.099,.124,.143,.158,.168,.174
*     LEAF ANGLE DISTRIBUTION, NOT CUMULATIVE, SUMMING TO UNITY
      PI =4.*ATAN(1.)
      RAD=PI/180.

***** 1. EXPOSITION OF LEAVES TO THE SUN

PROCEDURE SUMF,ZISSN=GEOMET(RAD)
      SUMF=F(1)+F(2)+F(3)+F(4)+F(5)+F(6)+F(7)+F(8)+F(9)
      IF (SUMF.NE.0.) GO TO 10
*     WHEN NO LEAF DISTRIBUTION FUNCTION IS PROVIDED
*     A SPHERICAL LEAF ANGLE DISTRIBUTION IS ASSUMED
      ZISSN=0.1
        DO 11 IS=1,9
   11   DAV(IS)=0.5
      GO TO 24
   10   DO 23 IS=1,9
        FI=(10*IS-5)*RAD
        SI=SIN(FI)
        CO=COS(FI)
        DD=0.
          DO 19 IL=1,9
          FL=(10*IL-5)*RAD
          AA=SI*COS(FL)
          BB=CO*SIN(FL)
          CC=AA
          IF (IS.GE.IL) GO TO 14
          SQ=SQRT(BB*BB-AA*AA)
          CC=2.*(AA*ATAN(AA/SQ)+SQ)/PI
   14     DD=DD+CC*F(IL)
            DO 18 SN=1,9
            FN=SN/10.
            FA=FN-AA
            CC=1.
            IF (IS.LT.IL) GO TO 15
            IF (FN-BB.GE.AA) GO TO 18
            IF (FN+BB.GT.AA) GO TO 17
            CC=0.
            GO TO 18
   17       SQ=SQRT(BB*BB-FA*FA)
            CC=ATAN(FA/SQ)/PI+0.5
            GO TO 18
   15       IF (FN-AA.GE.BB) GO TO 18
            IF (FN+AA.GE.BB) GO TO 17
            SQ=SQRT(BB*BB-FA*FA)
```

```
            CC=ATAN(FA/SQ)
            FA=FN+AA
            SQ=SQRT(BB*BB-FA*FA)
            CC=(ATAN(FA/SQ)+CC)/PI
   18       S(IL,SN)=CC
   19     S(IL,10)=1.
       EE=0.
         DO 22 SN=1,10
         CC=0.
            DO 21 IL=1,9
   21       CC=CC+F(IL)*S(IL,SN)
         Z(IS,SN)=CC-EE
   22    EE=CC
   23  DAV(IS)=DD
   24 CONTINUE
ENDPRO

***** 2. REFLECTION AND EXTINCTION

PROCEDURE EDIFB,EDIFV,EDIFN,KBL,KDFV,KDFN=EXTINC(SUMF)
STORAGE B(9), REV(11), RFN(11), KDN(11), KDV(11), KDIR(11)
TABLE B(1-9)=.030,.087,.133,.163,.174,.163,.133,.087,.030
*       DISTRIBUTION OF INCIDENT FLUXES OVER 9 ZONES OF THE SKY (UOC)
        SQVI=SQRT(1.-SCV)
PARAM SCV =0.2
*       SCATTERING COEFFICIENT OF THE LEAVES IN VISIBLE REGION
        SQNI=SQRT(1.-SCN)
PARAM SCN =0.85
*       SCATTERING COEFFICIENT OF THE LEAVES IN NEAR-INFRARED REGION
        REFV=(1.-SQVI)/(1.+SQVI)
        REFN=(1.-SQNI)/(1.+SQNI)
          DO 25 IS=1,9
          KDIR(IS+1)=DAV(IS)/SIN( (10*IS-5)*RAD)
          KDN (IS+1)=KDIR(IS+1)*SQNI*0.94623+0.03533
   25     KDV (IS+1)=KDIR(IS+1)*SQVI*0.94623+0.03533
        KDIR( 1)=KDIR( 2)
        KDIR(11)=KDIR(10)
        KDV ( 1)=KDV ( 2)
        KDV (11)=KDV (10)
        KDN ( 1)=KDN ( 2)
        KDN (11)=KDN (10)

*       DIRECT RADIATION
          DO 26 IS=1,11
          KDIRIS =2.*KDIR(IS)/(KDIR(IS)+1.)
          RFV(IS)=AMAX1(0.,1.117*(1.-EXP(-REFV*KDIRIS) )-0.0111)
   26     RFN(IS)=AMAX1(0.,1.117*(1.-EXP(-REFN*KDIRIS) )-0.0111)

*       DIFFUSE RADIATION
        RFOVV=0.
        RFOVN=0.
        EDIFB=0.
        EDIFV=0.
        EDIFN=0.
          DO 27 J=1,9
          RFOVV=RFOVV+B(J)*RFV(J+1)
          RFOVN=RFOVN+B(J)*RFN(J+1)
          EDIFB=EDIFB+B(J)*EXP(-KDIR(J+1) )
          EDIFV=EDIFV+B(J)*EXP(-KDV (J+1) )
   27     EDIFN=EDIFN+B(J)*EXP(-KDN (J+1) )
        KBL =-ALOG(EDIFB)
        KDFV=-ALOG(EDIFV)
        KDFN=-ALOG(EDIFN)
ENDPRO
```

```
***** 3. SITE AND STATE OF CROP

      CSLT=COS(RAD*LAT)
*     COSINE LATITUDE
      SNLT=SIN(RAD*LAT)
*     SINE LATITUDE
PARAM LAT =52.

      IWS=AFGEN(WSMTB,STDAY)
*     INITIAL WEIGHT SHOOT
      IWYR =600.*(1.-EXP(-IWS/4200.))
*     INITIAL WEIGHT OF YOUNG ROOT
      IWR  =IWYR+IWOR
*     INITIAL WEIGHT OF ROOTS
      IWOR =IWS/7.-IWYR
PARAM RESPI=0.03
      IRES =(IWS+IWR)*RESPI/(1.-RESPI)
*     INITIAL RESERVES OF THE PLANT

      IPS=AFGEN(FPTB,STDAY)*IWS
      ICS=AFGEN(FCTB,STDAY)*IWS
      IFS=AFGEN(FFTB,STDAY)*IWS
      ILS=AFGEN(FLTB,STDAY)*IWS
      IMS=AFGEN(FMTB,STDAY)*IWS
      IAS=AFGEN(FATB,STDAY)*IWS
      IPR=AFGEN(FPTB,STDAY)*IWR
      ICR=AFGEN(FCTB,STDAY)*IWR
      IFR=AFGEN(FFTB,STDAY)*IWR
      ILR=AFGEN(FLTB,STDAY)*IWR
      IMR=AFGEN(FMTB,STDAY)*IWR
      IAR=AFGEN(FATB,STDAY)*IWR
*     INITIAL AMOUNTS IN SHOOT AND ROOTS,RESPECTIVELY
*     PROTEINS, CARBOHYDRATES, LIPIDS, LIGNIN, MINERALS, ORGANIC ANIONS


      AVTCP=AFGEN(MNTT,STDAY)
*     AVERAGE TEMPERATURE OF CANOPY, INITIALLY

      IANCRL=IAMAS*1.629629/86400.
*     INITIAL VALUE OF FIRST ORDER AVERAGE OF CO2 ASSIMILATION
*     RATE (KG CO2/HA/S)
      IAMAS =AMIN1(IWS/3000.,1.)*200.
*     INITIAL AVERAGE METABOLIC ACTIVITY OF THE SHOOT, DEPENDENT ON
*     SHOOT WEIGHT (KG STARCH/HA/DAY)
      IAMAR =IWR/IWS*200.
*     INITIAL AVERAGE METABOLIC ACTIVITY IN THE ROOT (KG STARCH/HA/DAY)

DYNAMIC

************************     SECTION  3     ****************************

*        TIMER VARIABLES

      DAY =AMOD(AINT(TIME/86400.+STDAY),365.)
      HOUR=AMOD(TIME/3600.,24.)
PARAM STDAY=173.

*        CLIMATE

***** 1. WEATHER

PROCEDURE SNHSS,TA,TS,DPT,AVP,SVP,SLOPE,RH,WS=WEATH(DAY,HOUR)

*        DIRECTION OF THE SUN
      DEC  =-23.4*COS(2.*PI*(DAY+10.)/365.)
*     DECLINATION OF THE SUN
      SNDC =SIN(RAD*DEC)
```

```
*       SINE DECLINATION
        CSDC =COS(RAD*DEC)
*       COSINE DECLINATION
        SNHSS=SNLT*SNDC+CSLT*CSDC*COS(PI*(HOUR+12.)/12.)
*       SINE OF THE HEIGHT OF THE SUN
        LSNHS=INTGRL(-0.5,(SNHSS-LSNHS)/DELT)
*       SUN HEIGHT AT LAST TIME STEP
        RISE =ZHOLD(AND(SNHSS,-LSNHS)-0.5,HOUR-SNHSS*DELT/ ...
              ( (NOT(SNHSS-LSNHS)+SNHSS-LSNHS)*3600.)-RISEI)+RISEI
*       TIME OF SUN RISE TODAY, IN HOURS, ESTIMATE FOR TOMORROW
INCON RISEI=4.8

*          TEMPERATURE
        TA    =WAVE(DAY,HOUR,MNTT,MXTT,RISE)
*       TEMPERATURE AIR AT SCREEN HEIGHT
        TS    =INTGRL(20.,(TA-TS)/14400.)
*       TEMP. SOIL FOLLOWS AIR TEMP. WITH DELAY OF 4 HOURS

*          AIR HUMIDITY
        DPTC =WAVE(DAY,HOUR,EDPTTB,MDPTTB,RISE)
        DPT  =AMIN1(TA,DPTC)
*       CORRECTS FOR INPUT ERRORS LEADING TO DPT EXCEEDING TA
        AVP  =6.11*EXP(17.4*DPT/(239.+DPT) )
*       ACTUAL WATER VAPOUR PRESSURE, IN MBAR
        SVP  =6.11*EXP(17.4*TA/(239.+TA) )
*       SATURATION VAPOUR PRESSURE, IN MBAR
        SLOPE=4158.6*SVP/(TA+239.)**2
*       DERIVATIVE OF SATURATION PRESSURE WITH RESPECT TO TEMPERATURE
        RH   =AVP/SVP
*       RELATIVE HUMIDITY

*          WIND SPEED
        WS    =AFGEN(WSTB,DAY+HOUR/24.)* ...
               INSW(AND(HOUR-6.,18.-HOUR)-0.5,0.66667,1.33333)
*       WIND SPEED IN M/S, AT DAYTIME TWICE THAT AT NIGHT
ENDPRO

***** 2. RADIATION

PROCEDURE SNHS,DIFOV,DIFON,DIFCL,SUNDCL,CRC,CRO,DRC,DRO, ...
           FOV,FCL,LFOV,LFCL,CRAD,LWRI,IS=INRAD(SNHSS,TA,LAI)
        SNHS =AMAX1(0.,SNHSS)
        HSUN =ATAN(SNHS/SQRT(1.-SNHS*SNHS) )/RAD
        DIFOV=AFGEN(DFOVTB,HSUN)
*       DIFFUSE OVERCAST VISIBLE
        DIFON=0.7*DIFOV
*       DIFFUSE OVERCAST INFRARED
        DIFCL=AFGEN(DFCLTB,HSUN)
*       DIFFUSE CLEAR
        SUNDCL=AFGEN(SUNDTB,HSUN)
*       DIRECT CLEAR
        CRC  =(SUNDCL+DIFCL)*2.
*       CURRENT RADIATION CLEAR, ALL WAVELENGTHS
        CRO  = DIFOV +DIFON
*       CURRENT RADIATION OVERCAST
        DRC  =DLYTOT(DRCI,CRC)
        DRO  =DLYTOT(DROI,CRO)
INCON DROI =6.6E6, DRCI=3.3E7
        DRCP =ZHOLD(IMPULS(0.,86400.),DRC)
        DROP =ZHOLD(IMPULS(0.,86400.),DRO)
        DTR  =AFGEN(DTRT,DAY)*RADCV
        FCL  =(DTR-DROP)/(NOT(DRCP-DROP)+DRCP-DROP)
        FOV  =1.-FCL
        LFOV =LIMIT(0.,1.,FOV)
        LFCL =1.-LFOV
        CRAD =LFCL*CRC+LFOV*CRO
```

```
      LWRCI=-5.668E-8*(TA+273.)**4*(0.56-SQRT(0.75*AVP)*0.092)
*     SURFACE RADIATION AFTER BRUNT, 0.75 CONVERTS AVP TO MM HG
      ALWRC=LWRCI*(1.-EXP(-KBL*LAI) )
      LWROI=LWRCI*0.1
      ALWRO=ALWRC*0.1
      LWRI =LFOV*LWROI+LFCL*LWRCI
      ALWR =LFOV*ALWRO+LFCL*ALWRC
*     DISTRIBUTION
      IS   =(HSUN+10.)/10.
      FISUN=(HSUN+15.)/10.
      ISUN =FISUN
      FI   =FISUN-ISUN
      KDR  =KDIR(ISUN)*(1.-FI)+FI*KDIR(ISUN+1)
      KV   =KDV (ISUN)*(1.-FI)+FI*KDV (ISUN+1)
      KN   =KDN (ISUN)*(1.-FI)+FI*KDN (ISUN+1)
      RV   =RFV (ISUN)*(1.-FI)+FI*RFV (ISUN+1)
      RN   =RFN (ISUN)*(1.-FI)+FI*RFN (ISUN+1)
      ERI  =EXP(-KDR)
      EFRIV=EXP(-KV)
      EFRIN=EXP(-KN)
*     EXTINCTION OF ALL TYPES ASSUMED EXPONENTIAL
ENDPRO

************************       SECTION  4     ****************************

*         ENERGY BALANCE

PROCEDURE AVISO,ANIRO,AVISC,ANIRC,AVIS,ANETR,NETRS, ...
          TEHL,TSHL,AVTCP,NCRL=ENERGY(SRW,AMAX)
      RA     =PARL*SQRT(WDL/WS)*0.5
*     DIFFUSION RESISTANCE OF THE LAMINAR LAYER IN S/M,
*     FACTOR 0.5 ACCOUNTS FOR BOTH SIDES OF THE LEAF
PARAM PARL  =185.
PARAM WDL   =0.05
*     WIDTH OF THE LEAVES IN M
      IF (SNHS.GT.0.) GO TO 100

*     NIGHT PERIOD
  101 AVISO =0.
      ANIRO =0.
      AVISC =0.
      ANIRC =0.
      AVIS  =0.
      ANETR =ALWR
      NETRS =LWRI-ALWR
      EHL   =0.
      SHL   =ALWR/LAI
      TEHL  =0.
      TSHL  =ALWR
      AVTCP =TA+SHL*RA/RHOCP
PARAM RHOCP=1200.
*     HEAT CAPACITY OF THE AIR IN J/M**3 PER KELVIN
      NCRL  =-DPL*LAI/3600.
      GO TO 109

*     DAYLIGHT PERIOD
  100 TAVISO=(1.-RFOVV)*DIFOV
      TANIRO=(1.-RFOVN)*DIFON
      AVISO =TAVISO*(1.-EXP(-KDFV*LAI) )
      ANIRO =TANIRO*(1.-EXP(-KDFN*LAI) )
      TAVISC=TAVISO*DIFCL/DIFOV+SUNDCL*(1.-RV)
      TANIRC=TANIRO*DIFCL/DIFON+SUNDCL*(1.-RN)
      AVISC = AVISO*DIFCL/DIFOV+SUNDCL*(1.-RV)*(1.-EXP(-KV*LAI) )
      ANIRC = ANIRO*DIFCL/DIFON+SUNDCL*(1.-RN)*(1.-EXP(-KN*LAI) )
      ANETRO= AVISO+ ANIRO+ALWRO
      ANETRC= AVISC+ ANIRC+ALWRC
```

```
      TNETRO=TAVISO+TANIRO+LWROI
      TNETRC=TAVISC+TANIRC+LWRCI
      AVIS  =LFOV*AVISO +LFCL*AVISC
      ANETR =LFOV*ANETRO+LFCL*ANETRC
      TNETR =LFOV*TNETRO+LFCL*TNETRC
      NETRS =TNETR-ANETR
      TEHLO =0.
      TEHLC =0.
      TSHLO =0.
      TSHLC =0.
      AVTCPO=0.
      AVTCPC=0.
      NCRLO =0.
      NCRLC =0.
      IF (LAI.GT.0.2) GO TO 103

*     LAI NOT GREATER THAN 0.2
      RR    =RA/RHOCP
      DRYP  =(SVP-AVP)/RR
      LWO   =LWROI
      LWC   =LWRCI
      VISDFO=(1.-SCV)*DIFOV
      NIRDFO=(1.-SCN)*DIFON
      VISDFC=(1.-SCV)*(DIFCL+SUNDCL*KDR)
      NIRDFC=(1.-SCN)*(DIFCL+SUNDCL*KDR)
      TEHLO,TSHLO,AVTCPO,NCRLO=TRPH(VISDFO,NIRDFO,LWO,LAI)
      TEHLC,TSHLC,AVTCPC,NCRLC=TRPH(VISDFC,NIRDFC,LWC,LAI)
      GO TO 108

*     LAI GREATER THAN 0.2
  103 VISDFO=(1.-RFOVV)*(1.-EDIFV)*DIFOV
      NIRDFO=(1.-RFOVN)*(1.-EDIFN)*DIFON
      VIST  =SUNDCL*(1.-RV )*(1.-EFRIV)
      NIRT  =SUNDCL*(1.-RN )*(1.-EFRIN)
      VISD  =SUNDCL*(1.-SCV)*(1.-ERI  )
      NIRD  =SUNDCL*(1.-SCN)*(1.-ERI  )
      VISDFC=VISDFO*DIFCL/DIFOV+VIST-VISD
      NIRDFC=NIRDFO*DIFCL/DIFON+NIRT-NIRD
      LWO   =(1.-EDIFB)*LWROI
      LWC   =10.*LWO
      SUNPER=SUNDCL/SNHS
      SLLA  =(1.-ERI)/KDR
      SHLA  =1.-SLLA
      MAX   =LAI+1.
        DO 107 L=1,MAX
        LAIC  =L-1
        WSX   =AMAX1(WS*EXP(-0.7*LAIC),0.02)
        RA    =PARL*SQRT(WDL/WSX)*0.5
        RR    =RA/RHOCP
        DRYP  =(SVP-AVP)/RR
        LAIR  =LIMIT(0.01,1.,LAI-LAIC)
        IF (LAIR.EQ.1.) GO TO 104
  106   VISDFO=VISDFO*(1.-EXP(-KDFV*LAIR) )/(LAIR*(1.-EDIFV) )
        NIRDFO=NIRDFO*(1.-EXP(-KDFN*LAIR) )/(LAIR*(1.-EDIFN) )
        VIST  =VIST  *(1.-EXP(-KV   *LAIR) )/(LAIR*(1.-EFRIV) )
        NIRT  =NIRT  *(1.-EXP(-KN   *LAIR) )/(LAIR*(1.-EFRIN) )
        VISD  =VISD  *(1.-EXP(-KDR  *LAIR) )/(LAIR*(1.-ERI   ) )
        NIRD  =NIRD  *(1.-EXP(-KDR  *LAIR) )/(LAIR*(1.-ERI   ) )
        VISDFC=VISDFO*DIFCL/DIFOV+VIST-VISD
        NIRDFC=NIRDFO*DIFCL/DIFON+NIRT-NIRD
        LWO   =LWO*(1.-EXP(-KBL*LAIR) )/(LAIR*(1.-EDIFB) )
        LWC   =LWO*10.
        SLLA  =SLLA*(1.-EXP(-KDR*LAIR) )/(1.-ERI)
        SHLA  =LAIR-SLLA
  104   CONTINUE
        TEHLO,TSHLO,AVTCPO,NCRLO=TRPH(VISDFO,NIRDFO,LWO,LAIR)
```

```
          DO 105 SN=1,10
          IF (SUMF.NE.0.) ZISSN=Z(IS,SN)
          AREA=SLLA*ZISSN
          BLM =(0.1*SN-0.05)*SUNPER
          VIS =VISDFC+BLM*(1.-SCV)
          NIR =NIRDFC+BLM*(1.-SCN)
          TEHLC,TSHLC,AVTCPC,NCRLC=TRPH(VIS,NIR,LWC,AREA)
  105     CONTINUE
        TEHLC,TSHLC,AVTCPC,NCRLC=TRPH(VISDFC,NIRDFC,LWC,SHLA)
        IF (LAIR.LT.1.) GO TO 107
        VIST   =VIST*EFRIV
        NIRT   =NIRT*EFRIN
        VISD   =VISD*ERI
        NIRD   =NIRD*ERI
        VISDFO=VISDFO*EDIFV
        NIRDFO=NIRDFO*EDIFN
        VISDFC=VISDFO*DIFCL/DIFOV+VIST-VISD
        NIRDFC=NIRDFO*DIFCL/DIFON+NIRT-NIRD
        LWO    =LWO*EDIFB
        LWC    =LWC*EDIFB
        SLLA   =SLLA*ERI
        SHLA   =1.-SLLA
  107   CONTINUE

  108 TEHL  = LFOV*TEHLO +LFCL*TEHLC
      TSHL  = LFOV*TSHLO +LFCL*TSHLC
      AVTCP =(LFOV*AVTCPO+LFCL*AVTCPC)/LAI
      NCRL  =(LFOV*NCRLO +LFCL*NCRLC)/3600.
*     ENERGY FLUX (KG CO2/HA/S) AS MEASURED UNDER STANDARD CONDITIONS
  109 CONTINUE
ENDPRO
      DNETRS=DLYTOT(1000.,NETRS)
      DTLWR =DLYTOT(1000.,LWRI)
      DTABR =DLYTOT(1000.,ANETR)
*     BOWRAT=TSHL/(NOT(TEHL)+TEHL)


************************        SECTION  5      ***************************

*         WATER BALANCE

***** 1. BEGIN OF ITERATION

PROCEDURE RWCP,TEFS=BSCIL(TA,DIFOV,LAI,ACRS,SELECT)
      IF (TIME.EQ.0.) RWCPL=0.98
      RWCP=RWCPL
*     REL. WATER CONTENT OF CANOPY
      LOOP=0.
    1 TEFS=Q10**(0.1*AVTCP-2.5)
ENDPRO

***** 2. END OF ITERATION

PROCEDURE DIFF,LOOP=ESCIL(TRC,WUR,GRS)
      DIFF1=DIFF
      DIFF =TRC-WUR
*     DIFFERENCE BETWEEN TRANSPIRATION AND SUPPLY
      IF (LOOP.GT.0.) GO TO 2
      RWCP =RWCP-0.001
      GO TO 3
    2 IF (ABS(DIFF/WUR).LT.ERROR .OR. ABS(DIFF).LT.4.E-5) GO TO 4
PARAM ERROR=0.005
*     RELATIVE ERROR ALLOWED IN SELF-CONSTRUCTED IMPLICIT LOOP
      RWCLT=RWCP
      RWCP =RWCPL-(RWCP-RWCPL)*DIFF1/(DIFF-DIFF1)
      RWCPL=RWCLT
```

```
    3 LOOP =LOOP+1.
      IF (LOOP.LE.50.) GO TO 1
      IF (KEEP.GT.0.5) DIFF=101.
    4 RWCPL=RWCP
      TOTDIF=INTGRL(0.,DIFF)
ENDPRO

************************      SECTION  6      ****************************

*         CROP WATER STATUS

      SRW   =1./AFGEN(SRCTB,RWCP)
*     STOMATAL CONDUCTANCE DUE TO RELATIVE WATER CONTENT (M/S)
      TC    =TTRCC/(NOT(WSC-IWS)+WSC-IWS+RES-IRES)
*     TRANSPIRATION COEFFICIENT IN G WATER/G DRY MATTER
      DTTRC=DLYTOT(0.,TRC*1.E-3)
*     DAILY TOTAL OF WATER TRANSPIRED (KG/M**2, I.E., MM)
      TTRCC=INTGRL(0.,10.*TRC)
*     TOTAL WATER LOSS FROM THE CANOPY IN KG/HA
      TRC   =TEHL/VAPHT
*     TRANSPIRATION RATE (G/M**2/S), FROM TOTAL EVAPORATIVE HEAT
CONST VAPHT=2390.

      TWUR =INTGRL(0.,WUR*1.E4)
      WUR   =(WPTSL-WPTC)*ACRS
*     WATER UPTAKE BY THE ROOTS IN G/M**2/S
PARAM WPTSL=-0.1
*     WATER POTENTIAL SOIL, 0.1 CORRESPONDS WITH FIELD CAPACITY.
      WPTC =AFGEN(WPTTB,RWCP)
*     WATER POTENTIAL CROP IN BAR
      ACRS =(WYR+0.3*WDR)*AFGEN(ETRCTB,TS)/WCRR
*     ACTUAL CONDUCTANCE OF ROOT SYSTEM, G WATER/M**2/BAR/S
PARAM WCRR =2500.
*     WEIGHT/CONDUCTANCE RATIO OF ROOT SYSTEM


************************      SECTION  7      ****************************


***** 1. PHOTOSYNTHESIS

PARAM EFF=0.50
*     EFFICIENCY AT LOW LIGHT INTENSITIES, KG CO2/HA/HOUR PER J/M**2/S
      RMES =(RCO2IM-CO2C)*68.4/(AMAX1(0.001,AM300) *REDFRL)
*     AMAX1 PREVENTS DIVIDE CHECK
      REDFRL=AFGEN(REDFRT,RESL)
*     REDUCTION FACTOR ACCOUNTING FOR FEEDBACK OF RESERVE LEVEL
*     TO PHOTOSYNTHESIS
FUNCTION REDFRT=0.,1., .20,1., .25,.0001, 1.,.0001
      AMAX =(RCO2I-CO2C)*68.4/RMES
*     MAXIMUM CO2-ASSIMILATION OF SINGLE LEAVES
PARAM CO2C =10.
*     CO2 COMPENSATIONPOINT
      RCO2I=AMIN1(RCO2IM,RIECO2*ECO2C)
*     INTERNAL CO2-CONCENTRATION MAXIMALLY 120 VPPM FOR C4-PLANTS
*     AND 210 VPPM FOR C3-PLANTS
PARAM RCO2IM=120., RIECO2=0.6
      AM300=AFGEN(AMTB,TA)
FUNCTION AMTB=0.,0., 8.,0., 13.,70., 100.,70.

      DPL   =0.11*ANCRL/LAI*3600.
*     DISSIMILATION IN PHOTOSYNTHESIZING LEAVES (KG CO2/HA LEAF/S)
*     USED IN STOMATA REGULATION
      ANCRL=INTGRL(IANCRL,(AMAX1(0.,NCRL)-ANCRL)/(43200.))
*     FIRST ORDER AVERAGE OF NET CO2-ASSIMILATION
```

```
      NCASC=NCAS*3600.
*     NET CO2-ASSIMILATION OF SHOOT CALCULATED IN KG CO2/HA/H
      NCAS =NCRL-GRS-MRS
*     NET CO2-ASSIMILATION RATE OF SHOOT (KG CO2/HA/S)

***** 2. RESERVES

      RESL =RES/(RES+WSC+WRC)
*     RESERVE LEVEL
      RES  =INTGRL(IRES,NCRL/1.629629-URES)
*     AMOUNT OF RESERVES (KG STARCH/HA).
      URES =SR+SS+SRSOA+SRDOA
*     USE OF RESERVES
      SELECT=RESL*(FPR+FCR+FFR+FLR+FMR*SRMR)
*     SELECTION FORMULA TO ASSIST ITERATION

***** 3. EFFICIENCY OF GROWTH

*         SHOOT

      SS    =SRGS+SRMS+SRGSAS+SRTPAS
*     STARCH REQUIREMENT OF THE SHOOT (KG STARCH/HA/S)
      SRGS  =SRGPS+SRGCS+SRGFS+SRGMS+SRGLS
*     STARCH REQUIREMENT FOR GROWTH OF THE SHOOT (KG STARCH/HA/S)
      SRGPS =GRPS*0.517909
      SRGCS =GRCS*1.12
      SRGFS =GRFS*2.73
      SRGLS =GRLS*1.94
      SRGMS =GRMS*0.0184
*     STARCH REQUIREMENT FOR CONVERSION AND TRANSLOCATION OF
*     PROTEINS,CARBOHYDRATES,FATS,LIGNIN AND MINERALS IN THE
*     SHOOT (KG STARCH/HA/S)
      SRGSAS=GRPS*1.25
*     STARCH REQUIREMENT FOR FORMATION OF SKELETONS OF AMINO ACIDS
*     FOR SHOOT PROTEINS (KG STARCH/HA/S)
      SRTPAS=GS*FAS*0.0391
*     STARCH REQUIREMENT FOR TRANSPORT OF ORGANIC ANIONS IN THE
*     SHOOT (KG STARCH/HA/S)

*         ROOT

      SRGPR =GRPR*0.517909
      SRGCR =GRCR*1.12
      SRGFR =GRFR*2.73
      SRGLR =GRLR*1.94
      SRGMR =GRMR*0.0184
*     STARCH REQUIREMENT FOR CONVERSION AND TRANSLOCATION OF
*     PROTEINS,CARBOHYDRATES,FATS,LIGNIN AND MINERALS IN THE
*     ROOT (KG STARCH/HA/S)
      SR    =SRGR+SRMR+SRGSAR+SMU+SRTPAR
*     TOTAL STARCH REQUIREMENT OF THE ROOT (KG STARCH/HA/S)
      SRGR  =SRGPR+SRGCR+SRGFR+SRGLR+SRGMR
*     STARCH REQUIREMENT FOR GROWTH OF THE ROOT
      SRGSAR=GRPR*1.25
*     STARCH REQUIREMENT FOR THE FORMATION OF SKELETONS OF
*     AMINO ACIDS FOR ROOT PROTEINS (KG STARCH/HA/S)
      SRTPAR=GRAR*.0391
*     STARCH REQUIREMENT FOR TRANSPORT OF ORGANIC ANIONS IN THE ROOT
      SMU   =0.035*(GRMS+GRMR+RNO3)
*     STARCH REQUIREMENT FOR THE UPTAKE OF MINERALS AND NITRATE
*     (KG STARCH/HA/S)

      SRSOA =(RAF-TRRT)*0.8916
*     STARCH REQUIREMENT FOR FORMATION OF SKELETONS OF ORGANIC ANIONS
      SRDOA =TRRT*0.366341
*     STARCH LOST BY CO2 EVOLUTION DURING DECARBOXYLATION
*     OF ORGANIC ANIONS
```

```
***** 4. MAINTENANCE

      SRMS  =((PS*0.0225+MS*0.03)*FNATS+AMAS*0.04)*TEFS/86400.
*     STARCH REQUIREMENT FOR MAINTENANCE OF THE SHOOT (KG STARCH/HA/S)
      FNATS =1.-AMIN1(1.,750.*LAI/WSC)
      AMAS  =INTGRL(IAMAS,(SS-AMAS/86400.)*2.)
*     AVERAGE METABOLIC ACTIVITY SHOOT (KG STARCH PRODUCED AND USED/DAY)
*     TEFS =Q10**(0.1*AVTCP-2.5)  TEMP. EFFECT IN SHOOT
PARAM Q10  =2.0
      SRMR  =(PR*0.0225+MR*0.03+AMAR*0.04)*TEFR/86400.
*     STARCH REQUIREMENT FOR MAINTENANCE OF THE ROOT (KG STARCH/HA/S)
      AMAR  =INTGRL(IAMAR,(SR-AMAR/86400.)*2.)
      TEFR  =Q10**(0.1*TS-2.5)

***** 5. CO2 EVOLUTION

      GRS   =GRPS*0.844+GRCS*0.175+GRFS*1.618+GRLS*0.62+GRMS*0.03...
              +SRTPAS*1.629629
*     CO2 EVOLUTION RESULTING FROM GROWTH OF THE SHOOT (KG CO2/HA/S)
      MRS   =SRMS*1.629629
*     CO2 EVOLUTION RESULTING FROM MAINTENANCE OF THE SHOOT(KG CO2/HA/S)
      DR    =GRR+MRR+RUPT+DDA
*     TOTAL CO2 EVOLUTION IN THE ROOT (KG CO2/HA/S)
      GRR   =GRPR*0.844+GRCR*0.175+GRFR*1.618+GRLR*0.62+GRMR*0.03...
              +SRTPAR*1.629629
*     CO2 EVOLUTION RESULTING FROM GROWTH OF THE ROOT (KG CO2/HA/S)
      MRR   =SRMR*1.629629
*     CO2 EVOLUTION RESULTING FROM MAINTENANCE OF THE ROOT(KG CO2/HA/S)
      RUPT  =SMU*1.629629
*     CO2 EVOLUTION RESULTING FROM THE UPTAKE OF MINERALS
      DDA   =TRRT*.597
*     CO2 EVOLUTION RESULTING FROM DECARBOXYLATION OF ORGANIC ANIONS

***** 6. CARBON BALANCE

      RDPF =(WCP-WCF)/(NOT(WCP)+WCP)
      WCP  =((PS-IPS)+(PR-IPR))*.55555+ ...
             ((CS-ICS)+(CR-ICR))*.45005+ ...
             ((FS-IFS)+(FR-IFR))*.77206+ ...
             ((LS-ILS)+(LR-ILR))*.69313+ ...
             ((AS-IAS)+(AR-IAR))*.39627
      WCF  =TNCAP*.272727+(IRES-RES)*.444444
*     WCP AND WCF IN KG CARBON/HA, THE DIFFERENCE BETWEEN BOTH MUST BE
*     SMALLER THAN 0.01 TIMES THEIR VALUE
      TNCAP=INTGRL(0.,NCAS-DR)
*     TOTAL NET CO2-ASSIMILATION OF PLANT (KG CO2/HA)


************************      SECTION  8      ***************************


***** 1. CROP GROWTH

      TWT  =WSC+WRC+RES
*     TOTAL DRY WEIGHT (KG/HA)
      TWS  =WSC+RES
*     TOTAL WEIGHT SHOOT

      WSM  =AFGEN(WSMTB,DAY+HOUR/24.)
*     WEIGHT OF SHOOT MEASURED
      WSN  =AFGEN(WSMTB,DAY+1.)
      WSO  =AFGEN(WSMTB,DAY)
      DTGSM=DLYTOT(0.,GSM)
*     DAILY TOTAL OF GROWTH OF SHOOT AS MEASURED.
      GSM  =(AFGEN(WSMTB,DAY+0.5)-AFGEN(WSMTB,DAY-0.5) )/86400.
*     GAIN OF SHOOT WEIGHT MEASURED, IN KG DRY MATTER/HA/S
```

```
      DTGSC=DLYTOT(0.,GS)
*     DAILY TOTAL OF INCREASE IN STRUCTURAL WEIGHT OF THE SHOOT
      LAI  =AFGEN(LAITB,DAY+HOUR/24.)
*     LEAF AREA INDEX IN M**2/M**2

***** 2. SHOOT

      WSC  =PS+CS+FS+LS+MS+AS
*     WEIGHT SHOOT CALCULATED

      PS   =INTGRL(IPS,GRPS)
      CS   =INTGRL(ICS,GRCS)
      FS   =INTGRL(IFS,GRFS)
      LS   =INTGRL(ILS,GRLS)
      MS   =INTGRL(IMS,GRMS)
      AS   =INTGRL(IAS,GRAS)
*     WEIGHT OF PROTEINS,CARBOHYDRATES,FATS,LIGNIN,MINERALS AND
*     ORGANIC ANIONS IN THE SHOOT (KG/HA)

      GRPS =GS*FPS
      GRCS =GS*FCS
      GRFS =GS*FFS
      GRLS =GS*FLS
      GRMS =GS*FMS
      GRAS =GS*FAS
*     GROWTH RATE OF PROTEINS,CARBOHYDRATES,FATS,LIGNIN,MINERALS AND
*     ORGANIC ANIONS IN THE SHOOT (KG/HA/S)

*     IN PHOTON FPS,FCS,FFS,FLS,FMS AND FAS ARE GIVEN AS PARAMETERS
*     CHEMICAL COMPOSITION DOES NOT CHANGE DURING CALCULATION

      FPS=INCREM(FPTB)
      FCS=INCREM(FCTB)
      FFS=INCREM(FFTB)
      FLS=INCREM(FLTB)
      FMS=INCREM(FMTB)
      FAS=INCREM(FATB)
*     CHEMICAL COMPOSITION OF MATERIAL GROWING CURRENTLY (FRACTIONS)

      GS   =RES*RCRS*AFGEN(TGTB,AVTCP)*AFGEN(WGSTB,RWCP)
*     GROWTH RATE OF THE SHOOT (KG DM/HA/S)
PARAM RCRS=1.3E-5
*     RELATIVE CONSUMPTION RATE OF THE RESERVES

*     TGTB REPRESENTS INFLUENCE OF TEMPERATURE ON GROWTH
*     (SPECIES DEPENDENT)
*     WGSTB ACCOUNTS FOR CHANGES IN PARTITIONING OF RESERVES BETWEEN
*     SHOOT AND ROOT, UNDER INFLUENCE OF WATER CONTENT

      RAF  =RNO3*1.06
*     RATE OF FORMATION OF ORGANIC ANIONS,CONCURRENT WITH
*     RATE OF NITRATE REDUCTION
      RNO3 =(GS*FPS+GYR*FPR)*0.652
*     RATE OF NITRATE REDUCTION, ASSUMING THAT ALL PROTEIN N ORIGINATES
*     FROM NITRATES (KG NO3/HA/S)
      TRRT =RAF-GS*FAS-GYR*FAR
*     RATE OF TRANSPORT OF ORGANIC ANIONS TO THE ROOT FOR
*     DECARBOXYLATION (KG/HA/S)

***** 3. ROOT

      WRC  =WOR+WYR
*     WEIGHT OF ROOTS CALCULATED
      WOR  =INTGRL(IWOR,SYR-WOR/ROTC)
*     WEIGHT OF OLD ROOTS (SUBERIZED)
      SYR  =WYR*AFGEN(ETRCTB,TS)/SUBC
```

```
*     SUBERIZATION RATE OF YOUNG ROOTS (KG/HA/S)
PARAM SUBC =4.50E5
*     TIME CONSTANT FOR SUBERIZATION OF YOUNG ROOTS (= 5 DAYS)
*     ETRCTB ACCOUNTS FOR THE INFLUENCE OF TEMPERATURE ON THE RATE
*     OF SUBERIZATION
PARAM ROTC =30.E6
*     TIME CONSTANT FOR ROTTING OF OLD ROOTS (ABOUT ONE YEAR, SO
*     INEFFECTIVE HERE)
      WYR  =INTGRL(IWYR,GYR-SYR)
*     WEIGHT OF YOUNG ROOTS (KG DM/HA)
      GYR  =RES*RCRS*AFGEN(TGTB,TS)*AFGEN(WGRTB,RWCP)
*     GROWTH RATE OF YOUNG ROOTS (KG DM/HA/S)
*     WGRTB ACCOUNTS FOR THE INFLUENCE OF WATER CONTENT ON THE
*     PARTITIONING BETWEEN SHOOT AND ROOT

      PR   =INTGRL(IPR,GRPR)
      CR   =INTGRL(ICR,GRCR)
      FR   =INTGRL(IFR,GRFR)
      LR   =INTGRL(ILR,GRLR)
      MR   =INTGRL(IMR,GRMR)
      AR   =INTGRL(IAR,GRAR)
*     AMOUNT OF PROTEINS,CARBOHYDRATES,FATS,LIGNIN,MINERALS AND
*     ORGANIC ANIONS IN THE ROOT (KG DM/HA)

      GRPR  =GYR*FPR
      GRCR  =GYR*FCR
      GRFR  =GYR*FFR
      GRLR  =GYR*FLR
      GRMR  =GYR*FMR
      GRAR  =GYR*FAR
*     GROWTH RATE OF PROTEINS,CARBOHYDRATES,FATS,LIGNIN,MINERALS AND
*     ORGANIC ANIONS IN THE ROOT (KG DM/HA/S)

      FPR=INSW(FPS-0.1,FPS,FPS-0.02)
      FCR=INSW(FPS-0.1,FCS,FCS+0.02)
      FFR=FFS
      FLR=FLS
      FMR=FMS
      FAR=FAS
*     CHEMICAL COMPOSITION OF ROOT MATERIAL GROWING, FRACTION OF
*     PROTEINS,CARBOHYDRATES,FATS,LIGNIN,MINERALS AND ORGANIC ANIONS
*     DERIVED FROM THOSE IN THE SHOOT

************************      SECTION  9      ***************************

***** 1. ACTUAL PLANT DATA

FUNCTION AMTB=0.,0., 8.,0.,13.,70., 100.,70.

FUNCTION SRCTB=.7,1.E-4, .84,1.E-4, .95,1.E-4, 1.,.01429, 1.5,.01429
*     DEPENDENCE OF STOMATAL CONDUCTANCE (M/S) ON RWCP, FROM
*     MAIZE PRETREATED AT HIGH INTENSITIES IN PHOTOSYNTHESIS ROOM
FUNCTION WPTTB=-1.,-200., 0.5,-50., 0.7,-17., 0.8,-14., 0.84,-12.5, ...
               0.88,-10., 0.9,-8.1, 1.,0., 1.5,40.5, 2.5,200.
*     DEPENDENCE OF WATER POTENTIAL CROP (BAR) ON RWCP
FUNCTION ETRCTB=0.,0., 10.,0.08, 20.,0.29, 33.,0.94, 37.,1.
*     EFFECT OF SOIL TEMPERATURE ON ROOT CONDUCTANCE

FUNCTION WSMTB =173.,48.,   187.,217.,  201.,1082., 208.,3082., ...
                220.,5681., 229.,6647., 241.,10013.,255.,12012.
*     DRY MATTER 1972 MAIZE FLEVOLAND
FUNCTION LAITB =173.,0.086, 187.,0.409, 201.,2.034, 208.,3.943, ...
                220.,3.876, 229.,3.429, 241.,3.444, 255.,3.2
*     LAI 1972 MAIZE FLEVOLAND

FUNCTION FPTB  =173.,.270,  187.,.270,  201.,.230,  208.,.220,  ...
```

```
               220.,.165,  229.,.150,  241.,.130,  255.,.110
          CRUDE PROTEIN 1972, MAIZE FLEVOLAND
FUNCTION FCTB  =173.,.494,  187.,.494,  201.,.528,  208.,.531,  ...
               220.,.581,  229.,.606,  241.,.634,  255.,.661
*         CARBOHYDRATE 1972, MAIZE FLEVOLAND
FUNCTION FFTB  =173.,.025,  255.,.025
*         FAT 1972, MAIZE FLEVOLAND
FUNCTION FLTB  =173.,.050,  187.,.050,  201.,.070,  208.,.080,  ...
               220.,.100,  255.,.100
*         LIGNIN 1972 ESTIMATED, MAIZE FLEVOLAND
FUNCTION FMTB  =173.,.065,  255.,.065
*         MINERAL 1972, MAIZE FLEVOLAND
FUNCTION FATB  =173.,.096,  187.,.096,  201.,.082,  208.,.079,  ...
               220.,.064,  229.,.054,  241.,.046,  255.,.039
*         ORGANIC ANIONS 1972, MAIZE FLEVOLAND

FUNCTION WGRTB=0.,1., .8,1., .9,.85,.975,.5, 1.,0.
*      RELATION BETWEEN RELATIVE WATER CONTENT AND GROWTH RATE OF ROOT
FUNCTION WGSTB=0.,0., .8,0., .9,.15, .975,.5, 1.,1.
*      RELATION BETWEEN RELATIVE WATER CONTENT AND GROWTH OF THE SHOOT
FUNCTION TGTB=0.,0., 10.,0., 25.,1., 35.,1., 40.,0.
*      RELATION BETWEEN TEMPERATURE AND GROWTH


***** 2. METEOROLOGICAL DATA

FUNCTION DFOVTB=0.,0., 5.,6., 15.,26., 25.,45., 35.,64., 45.,80., ...
               55.,94., 65.,105., 75.,112., 90.,116.
FUNCTION DFCLTB=0.,0., 5.,29., 15.,42., 25.,49., 35.,56., 45.,64., ...
               55.,68., 65.,71., 75.,75., 90.,77.
FUNCTION SUNDTB=0.,0., 5.,0., 15.,88., 25.,175., 35.,262., 45.,336., ...
               55.,402.,65.,452., 75.,483., 90.,504.
*         RADIATION VALUES FOR STANDARD SKIES


FUNCTION WSTB = 140.,3.,141.,4.,142.,4.,143.,3.,144.,3.,...
               145.,6.,146.,8.,147.,13,148.,10,149.,9.,...
               150.,7.,151.,8.,152.,7.,153.,3.,154.,2.,...
               155.,4.,156.,4.,157.,4.,158.,3.,159.,4.,...
               160.,6.,161.,4.,162.,3.,163.,4.,164.,3.,...
               165.,3.,166.,4.,167.,4.,168.,4.,169.,2.,...
               170.,6.,171.,6.,172.,4.,173.,7.,174.,5.,...
               175.,6.,176.,3.,177.,4.,178.,3.,179.,4.,...
               180.,4.,181.,3.,182.,3.,183.,4.,184.,5.,...
               185.,3.,186.,5.,187.,3.,188.,2.,189.,4.,...
               190.,4.,191.,3.,192.,4.,193.,3.,194.,2.,...
               195.,2.,196.,2.,197.,3.,198.,5.,199.,5.,...
               200.,4.,201.,3.,202.,3.,203.,3.,204.,4.,...
               205.,2.,206.,3.,207.,3.,208.,5.,209.,6.,...
               210.,4.,211.,2.,212.,2.,213.,4.,214.,3.,...
               215.,2.,216.,5.,217.,5.,218.,5.,219.,4.,...
               220.,3.,221.,5.,222.,4.,223.,3.,224.,4.,...
               225.,4.,226.,3.,227.,3.,228.,4.,229.,3.,...
               230.,5.,231.,5.,232.,5.,233.,3.,234.,5.,...
               235.,5.,236.,2.,237.,4.,238.,4.,239.,2.,...
               240.,2.,241.,4.,242.,4.,243.,5.,244.,5.,...
               245.,3.,246.,4.,247.,4.,248.,3.,249.,2.,...
               250.,2.,251.,3.,252.,2.,253.,6.,254.,5.,...
               255.,5.,256.,3.,257.,2.,258.,3.,259.,3.,...
               260.,5.,261.,4.,262.,2.,263.,3.,264.,2.
*         WIND SPEED 1972

FUNCTION MDPTTB=140., 2.3,                                        ...
               141., 8.1,142., 5.4,143., 6.8,144.,10.7,145.,11.1,...
               146., 6.4,147., 4.5,148., 8.8,149., 8.1,150., 6.2,...
               151., 6.0,152.,10.7,153., 6.8,154., 7.2,155.,11.1,...
```

```
               156.,12.0,157.,13.8,158., 7.8,159., 7.8,160., 6.6,...
               161., 9.4,162., 7.4,163., 7.8,164., 7.6,165., 8.7,...
                      167., 9.2,168., 9.9,169., 6.4,170.,10.5,...
               171., 5.6,172., 5.8,173.,11.3,174.,11.1,175., 7.2,...
               176.,10.4,177.,16.4,178.,13.0,179.,16.4,180., 9.2,...
               181., 7.4,182., 9.9,183.,11.0,184.,10.4,185.,12.0,...
               186.,13.7,187.,15.3,188.,14.8,189.,10.8,190.,14.5,...
               191.,15.6,192., 6.4,193.,13.2,194., 6.0,195.,12.0,...
               196.,12.5,197.,13.2,198.,16.6,199.,20.3,200.,22.1,...
               201.,22.1,202.,20.9,203.,17.9,204.,20.3,205.,15.8,...
               206.,18.8,207.,13.6,208.,10.2,209.,15.9,210.,13.9,...
               211.,13.0,212.,13.3,213.,11.7,214.,12.6,215.,14.3,...
               216.,12.7,217.,10.2,218.,12.9,219.,14.1,220.,15.6,...
               221.,16.9,222.,10.4,223.,12.5,224.,10.0,225., 9.0,...
               226.,17.1,227.,13.2,228.,12.7,229., 8.7,230.,13.4,...
               231., 8.8,232.,10.4,233.,10.5,234.,11.9,235., 9.9,...
               236., 9.4,237.,13.0,238.,14.5,239.,10.4,240.,12.5,...
               241., 9.2,242., 8.5,243., 8.5,244., 9.4,245., 8.3,...
               246., 8.1,247.,11.0,248.,10.7,249., 8.7,250.,11.6,...
               251.,13.9,252., 9.7,253.,14.8,254., 9.2,255., 8.1,...
               256., 9.7,257., 6.4,258., 7.0,259., 6.2,260., 8.0,...
               261., 6.8,262., 8.8,263., 9.2,264., 8.7,265., 7.0
*          DEW POINT 1972

FUNCTION EDPTTB=140., 2.3,                                          ...
               141., 8.1,142., 5.4,143., 6.8,144.,10.7,145.,11.1,...
               146., 6.4,147., 4.5,148., 8.8,149., 8.1,150., 6.2,...
               151., 6.0,152.,10.7,153., 6.8,154., 7.2,155.,11.1,...
               156.,12.0,157.,13.8,158., 7.8,159., 7.8,160., 6.6,...
               161., 9.4,162., 7.4,163., 7.8,164., 7.6,165., 8.7,...
                      167., 9.2,168., 9.9,169., 6.4,170.,10.5,...
               171., 5.6,172., 5.8,173.,11.3,174.,11.1,175., 7.2,...
               176.,10.4,177.,16.4,178.,13.0,179.,16.4,180., 9.2,...
               181., 7.4,182., 9.9,183.,11.0,184.,10.4,185.,12.0,...
               186.,13.7,187.,15.3,188.,14.8,189.,10.8,190.,14.5,...
               191.,15.6,192., 6.4,193.,13.2,194., 6.0,195.,12.0,...
               196.,12.5,197.,13.2,198.,16.6,199.,20.3,200.,22.1,...
               201.,22.1,202.,20.9,203.,17.9,204.,20.3,205.,15.8,...
               206.,18.8,207.,13.6,208.,10.2,209.,15.9,210.,13.9,...
               211.,13.0,212.,13.3,213.,11.7,214.,12.6,215.,14.3,...
               216.,12.7,217.,10.2,218.,12.9,219.,14.1,220.,15.6,...
               221.,16.9,222.,10.4,223.,12.5,224.,10.0,225., 9.0,...
               226.,17.1,227.,13.2,228.,12.7,229., 8.7,230.,13.4,...
               231., 8.8,232.,10.4,233.,10.5,234.,11.9,235., 9.9,...
               236., 9.4,237.,13.0,238.,14.5,239.,10.4,240.,12.5,...
               241., 9.2,242., 8.5,243., 8.5,244., 9.4,245., 8.3,...
               246., 8.1,247.,11.0,248.,10.7,249., 8.7,250.,11.6,...
               251.,13.9,252., 9.7,253.,14.8,254., 9.2,255., 8.1,...
               256., 9.7,257., 6.4,258., 7.0,259., 6.2,260., 8.0,...
               261., 6.8,262., 8.8,263., 9.2,264., 8.7,265., 7.0
*          DEW POINT 1972 FLEVOLAND

FUNCTION MNTT = 140., 5.0,141., 5.1,142., 7.5,143., 7.5,144.,10.9,...
               145.,10.8,146.,11.8,147.,11.2,148., 8.9,149., 9.8,...
               150.,10.2,151., 9.0,152., 9.3,153., 7.5,154., 4.7,...
               155.,11.1,156.,12.2,157., 8.5,158.,11.1,159., 7.9,...
               160.,10.0,161., 7.5,162., 9.9,163., 7.8,164., 7.8,...
               165., 8.1,166., 9.8,167., 7.4,168., 8.9,169., 5.6,...
               170., 9.9,171.,12.0,172., 7.0,173.,10.0,174.,12.0,...
               175., 8.9,176.,10.0,177.,13.0,178., 9.7,179.,14.4,...
               180.,14.8,181.,12.8,182., 7.8,183.,10.1,184.,11.5,...
               185.,12.4,186., 8.0,187.,13.7,188.,16.0,189.,11.0,...
               190.,13.0,191.,14.5,192.,12.5,193.,10.1,194., 7.0,...
               195., 8.1,196., 8.8,197., 8.6,198.,14.8,199.,16.0,...
               200.,12.0,201.,18.9,202.,19.9,203.,19.2,204.,16.8,...
               205.,17.8,206.,15.8,207.,15.1,208.,14.0,209.,13.5,...
```

```
                      210.,14.8,211.,14.5,212.,14.9,213.,13.0,214.,12.7,...
                      215.,10.6,216.,13.0,217., 9.0,218.,13.3,219.,11.1,...
                      220.,14.9,221.,13.9,222.,11.9,223., 9.0,224.,14.7,...
                      225., 6.3,226.,15.0,227.,15.8,228.,14.9,229.,12.8,...
                      230.,10.0,231.,11.0,232.,10.6,233.,11.1,234.,13.5,...
                      235.,12.6,236., 8.9,237.,10.2,238.,11.1,239.,13.2,...
                      240., 8.0,241.,11.0,242.,10.2,243.,10.8,244.,10.5,...
                      245., 9.8,246., 7.9,247.,10.9,248., 8.0,249., 8.7,...
                      250., 6.9,251., 9.6,252.,12.0,253.,13.3,254.,11.4,...
                      255., 5.6,256., 8.1,257., 9.7,258., 8.7,259., 7.8,...
                      260., 8.2,261., 7.1,262., 6.5,263., 7.2,264., 4.8
          MIN TEMP 1972 FLEVOLAND


FUNCTION MXTT = 140.,14.3,141.,16.5,142.,17.9,143.,17.0,144.,22.0,...
                      145.,24.1,146.,14.1,147.,15.7,148.,16.1,149.,13.5,...
                      150.,13.9,151.,14.6,152.,15.1,153.,14.0,154.,14.9,...
                      155.,16.9,156.,15.3,157.,21.1,158.,21.2,159.,16.0,...
                      160.,20.0,161.,17.0,162.,17.1,163.,15.8,164.,16.1,...
                      165.,17.7,166.,18.2,167.,18.6,168.,17.8,169.,16.2,...
                      170.,20.0,171.,21.2,172.,17.2,173.,17.9,174.,17.0,...
                      175.,13.8,176.,15.0,177.,16.0,178.,20.1,179.,25.2,...
                      180.,24.3,181.,17.2,182.,17.9,183.,14.0,184.,14.8,...
                      185.,17.5,186.,17.8,187.,20.2,188.,26.9,189.,19.0,...
                      190.,19.0,191.,16.5,192.,17.3,193.,17.2,194.,18.5,...
                      195.,19.0,196.,19.9,197.,20.0,198.,21.7,199.,24.1,...
                      200.,24.2,201.,28.5,202.,30.0,203.,29.8,204.,29.8,...
                      205.,27.3,206.,21.1,207.,21.1,208.,19.0,209.,17.0,...
                      210.,18.0,211.,19.2,212.,19.2,213.,20.9,214.,19.0,...
                      215.,20.9,216.,19.1,217.,18.0,218.,18.8,219.,19.2,...
                      220.,24.1,221.,27.0,222.,24.2,223.,20.2,224.,21.1,...
                      225.,19.0,226.,22.2,227.,20.9,228.,19.8,229.,20.0,...
                      230.,18.0,231.,20.1,232.,16.5,233.,17.2,234.,18.5,...
                      235.,17.1,236.,17.1,237.,17.6,238.,18.8,239.,17.2,...
                      240.,19.1,241.,21.2,242.,21.2,243.,21.9,244.,21.5,...
                      245.,21.5,246.,19.7,247.,18.7,248.,19.8,249.,17.2,...
                      250.,18.0,251.,20.1,252.,21.1,253.,18.0,254.,21.5,...
                      255.,14.8,256.,15.8,257.,14.9,258.,15.0,259.,16.2,...
                      260.,14.9,261.,14.0,262.,15.0,263.,15.0,264.,19.2
*         MAX TEMP 1972 FLEVOLAND

FUNCTION DTRT = 140.,10034.,                                        ...
                      141.,10583.,142., 8223.,143.,10095.,144., 8811.,   ...
                      145., 2662.,146., 9088.,147., 8632.,148., 5504.,   ...
                      149., 5434.,150., 5473.,151., 6951.,152., 6598.,   ...
                      153., 8180.,154., 7898.,155., 4536.,156., 8510.,   ...
                      157., 6244.,158., 3196.,159., 8716.,160., 8344.,   ...
                      161., 7211.,162., 8403.,163., 8433.,164., 7938.,   ...
                      165., 9854.,166., 8720.,167., 7409.,168., 6710.,   ...
                      169., 9829.,170., 5945.,171., 9756.,172., 7865.,   ...
                      173., 4122.,174., 3188.,175., 6309.,176., 5322.,   ...
                      177., 6452.,178.,10001.,179., 6645.,180., 8618.,   ...
                      181., 8054.,182., 2518.,183., 2477.,184., 3506.,   ...
                      185., 7132.,186., 4808.,187., 9573.,188., 5032.,   ...
                      189., 7155.,190., 2379.,191., 2006.,192., 9228.,   ...
                      193., 7386.,194., 9491.,195.,10599.,196., 8532.,   ...
                      197.,10202.,198., 9833.,199., 6313.,200., 8191.,   ...
                      201., 8579.,202., 6547.,203., 8701.,204., 8431.,   ...
                      205., 5944.,206., 5360.,207., 4277.,208., 3547.,   ...
                      209., 3401.,210., 4244.,211., 5992.,212., 5837.,   ...
                      213., 6498.,214., 7714.,215., 4261.,216., 4048.,   ...
                      217., 5833.,218., 4536.,219., 9107.,220., 6727.,   ...
                      221., 4706.,222., 7280.,223., 7446.,224., 8385.,   ...
                      225., 8185.,226., 3140.,227., 5450.,228., 4920.,   ...
                      229., 6819.,230., 4163.,231., 6520.,232., 7917.,   ...
```

```
               233., 8141.,234., 6016.,235., 5730.,236., 5836.,    ...
               237., 5678.,238., 4581.,239., 4240.,240., 5151.,    ...
               241., 7770.,242., 7603.,243., 7864.,244., 7419.,    ...
               245., 6434.,246., 6570.,247., 6126.,248., 3711.,    ...
               249., 5050.,250., 6405.,251., 4785.,252., 3442.,    ...
               253., 3940.,254., 2939.,255., 4020.,256., 3061.,    ...
               257., 3663.,258., 4192.,259., 4700.,260., 3607.,    ...
               261., 3929.,262., 4471.,263., 4923.,264., 3263.,    ...
               265., 4822.,266., 3321.,267., 5898.,268., 5178.
*        RADIATION FLEVOLAND 1972
PARAM RADCV=2395.
*      RADIATION CONVERTED TO J/M**2

***** 3. OUTPUT

PRINT    AVISC,     ANETR,     DTLWR,     TEHL,     ...
         ANIRC,     NETRS,     DTABR,     TSHL,     ...
         AVTCP,     DNETRS,    FDV,       RWCP,     ...
         TA,        CRAD,      RH,        RDPF,     ...
         ANCRL,     TRC,       TC,        SRW,      ...
         MRR,       RND3,      DPL,       ACRS,     ...
         WRC,       GYR,       SR,        SRGS,     ...
         WSC,       GS,        SS,        SRMS,     ...
         RES,       WSM,       GRR,       DR,       ...
         NCAS,      HOUR,      WCP,       WCF,      ...
         NCASC,     DAY,       DTTRC,     TWS,      ...
         TTRCC,     WYR,       TWT

OUTPUT WSC,WSM
PAGE GROUP
***** 4. RUN CONTROL

TIMER DELT=3600., FINTIM=6998400., PRDEL=43200., OUTDEL=43200.
*      TIME IS EXPRESSED IN SECONDS
FINISH RDPF=0.01
METHOD RECT
END
STOP
ENDJOB
```

Appendix B – PHOTON: Simulation of daily photosynthesis and transpiration

```
TITLE SIMULATION OF DAILY PHOTOSYNTHESIS AND TRANSPIRATION

/      DIMENSION Z(9,10), S(9,10)
*  Z: DISTRIBUTION OF THE LEAVES WITH RESPECT TO INCOMING SUNRAYS

***** FOR EACH RUN THE FOLLOWING INPUT DATA ARE REQUIRED

*
***** MAIZE FLEVOLAND  1973
*
**            CLIMATE DATA
*      CRADTB INTENSITY OF SOLAR RADIATION IN CAL/CM**2/MIN
*      TATB   TEMPERATURE OF THE AIR IN DEGREES CENTIGRADE
*      DPTTB  DEW POINT TEMPERATURE, IN DEGREES CENTIGRADE
FUNCTION WSTB=0.,1., 1000.,1.
*             WIND SPEED IN THE ENCLOSURE, M/S

**            PHYSIOLOGICAL DATA
PARAM IWS= 12645.
*             SHOOT WEIGHT IN KG DRY MATTER/HA
PARAM RESPI= 0.03
*             INITIAL RESERVE PERCENTAGE, DEPENDING ON PRETREATMENT
PARAM FPS=.22, FCS=.53, FFS=.025, FLS=.08, FMS=.065, FAS=.08
PARAM FPR=.20, FCR=.55, FFR=.025, FLR=.08, FMR=.065, FAR=.08
*      CHEMICAL COMPOSITION OF PLANT MATERIAL GROWN ON THIS DATE
*      PROTEINS, CARBOHYDRATES, LIPIDS, LIGNIN, MINERALS, ORGANIC ANIONS
PARAM LAI= 5.3
*             LEAF AREA INDEX
*      F      LEAF ANGLE DISTRIBUTION
PARAM RESCW =2000.
*             CUTICULAR RESISTANCE TO WATER, IN S/M
FUNCTION SRCTB=.7,1.E-4, .84,1.E-4, .95,1.E-4,    1.,.01429, 1.5,.01429
*      DEPENDENCE OF STOMATAL CONDUCTANCE (M/S) ON RWCP, FROM
*      MAIZE PRETREATED AT HIGH INTENSITIES IN PHOTOSYNTHESIS ROOM
PARAM WDL    =0.05
*             WIDTH OF THE LEAVES IN M
FUNCTION TGTB=0.,0., 10.,0., 25.,1., 35.,1., 40.,0.
*             EFFECT OF TEMPERATURE ON GROWTH
FUNCTION WPTTB =.5,-50., .7,-17., .8,-14., .84,-12.5, .88,-10., ...
               .90,-8.1, 1.,0., 1.5,40.5
*      DEPENDENCE OF WATER POTENTIAL CROP (BAR) ON REL. WATER CONTENT
FUNCTION ETRCTB=0.,0., 10.,0.08, 20.,0.29, 33.,0.94, 37.,1.
*      EFFECT OF SOIL TEMPERATURE ON ROOT CONDUCTANCE,
*      FROM MAIZE TRANSPIRATION DATA IN PHOTOSYNTHESIS ROOM
*      TRCMTB TRANSPIRATION RATE IN G WATER/M2/H
PARAM CO2C =10.
PARAM RCO2IM=120., RIECO2=0.6
FUNCTION AMTB=0.,0., 8.,0., 13.,70., 100.,70.

**            ASSIMILATION CHAMBER DATA
*      VPMOTB CONCENTRATION OF THE OUTGOING CO2 IN VPPM
*      VPMITB CONCENTRATION OF THE INCOMING CO2 IN VPPM
*      LTAITB FLOW RATE OF THE AIR IN CHAMBER IN LITRES/H
PARAM AB=10000., AC=2.8
*      LENGTH OF SIDE OF ENCLOSURE, AND HEIGHT OF THE CROP,
```

```
*     WHEN ENCLOSURE IS SURROUNDED BY A CROP, AB SHOULD HAVE VALUE 1.E4
PARAM BAKOPP= 0.64
*              AREA OF THE ASSIMILATION CHAMBER

**             RUN CONTROL DATA
PARAM STDAY =226.
*              NUMBER OF CALENDAR DAY
PARAM BEGIN =11.
*              HOUR AT WHICH THE SIMULATION STARTS

*************************    SECTION  1    ********************************

*        AVERAGE TEMPERATURE OF CANOPY, SENSIBLE AND LATENT HEAT LOSS

MACRO TEHL,TSHL,AVTCP,NCRL,AVIS,ANIR=TRPH(VIS,NIR,AREA)
      ABSRAD=VIS+NIR
      EVA   =AMIN1(EFF*VIS/AMAX,46.)
*     PREVENTS UNDERFLOW
      NCRIL =(AMAX+DPL)*(1.-EXP(-EVA) )-DPL
      SRESL =(68.4*(VPMOC-RCO2I)/AMAX1(0.001,NCRIL)-RA*1.32)/1.66
      IF (SRESL.GT.SRW .OR. SRESL .LT. 0.) GO TO 700
      SRESL =SRW
      NCRIL =68.4*(VPMOC-RCO2I)/(SRW*1.66+RA*1.32)
  700 SRES  =AMIN1(RESCW,SRESL)
      ENP   =0.3*NCRIL
      EHL   =(SLOPE*(ABSRAD-ENP)+DRYP)/(PSCH*(RA*0.93+SRES)/RA+SLOPE)
      SHL   =ABSRAD-EHL-ENP
      TL    =TA+SHL*RR
      TEHL  =TEHL +AREA*EHL
      TSHL  =TSHL +AREA*SHL
      AVTCP =AVTCP+AREA*TL
      NCRL  =NCRL +AREA*NCRIL
      AVIS  =AVIS +AREA*VIS
      ANIR  =ANIR +AREA*NIR
ENDMAC
CONST PSCH=0.67
*     PSYCHROMETRIC CONSTANT IN MBAR PER KELVIN


*************************    SECTION  2    ********************************

*        INITIALIZATION
INITIAL
FIXED I,J,K,L,N,IL,IS,ISUN,SN
STORAGE F(9), DAV(9)
* DAV: AVERAGE PROJECTION OF THE LEAVES IN THE DIFFERENT DIRECTIONS
TABLE F(1-9)=.015,.045,.074,.099,.124,.143,.158,.168,.174
*     LEAF ANGLE DISTRIBUTION, NOT CUMULATIVE, SUMMING TO UNITY
      PI =4.*ATAN(1.)
      RAD=PI/180.

***** 1. EXPOSITION OF LEAVES TO THE SUN

PROCEDURE SUMF,ZISSN=GEOMET(RAD)
      SUMF=F(1)+F(2)+F(3)+F(4)+F(5)+F(6)+F(7)+F(8)+F(9)
      IF (SUMF.NE.0.) GO TO 10
*     WHEN NO LEAF DISTRIBUTION FUNCTION IS PROVIDED
*     A SPHERICAL LEAF ANGLE DISTRIBUTION IS ASSUMED
      ZISSN=0.1
        DO 12 IS=1,9
   12   DAV(IS)=0.5
      GO TO 11
   10   DO 20 IS=1,9
        FI=(10*IS-5)*RAD
        SI=SIN(FI)
        CO=COS(FI)
```

```
        DD=0.
          DO 21 IL=1,9
          FL=(10*IL-5)*RAD
          AA=SI*COS(FL)
          BB=CO*SIN(FL)
          CC=AA
          IF (IS.GE.IL) GO TO 22
          SQ=SQRT(BB*BB-AA*AA)
          CC=2.*(AA*ATAN(AA/SQ)+SQ)/PI
   22     DD=DD+CC*F(IL)
            DO 23 SN=1,9
            FN=SN/10.
            FA=FN-AA
            CC=1.
            IF (IS.LT.IL) GO TO 24
            IF (FN-BB.GE.AA) GO TO 23
            IF (FN+BB.GT.AA) GO TO 25
            CC=0.
            GO TO 23
   25       SQ=SQRT(BB*BB-FA*FA)
            CC=ATAN(FA/SQ)/PI+0.5
            GO TO 23
   24       IF (FN-AA.GE.BB) GO TO 23
            IF (FN+AA.GE.BB) GO TO 25
            SQ=SQRT(BB*BB-FA*FA)
            CC=ATAN(FA/SQ)
            FA=FN+AA
            SQ=SQRT(BB*BB-FA*FA)
            CC=(ATAN(FA/SQ)+CC)/PI
   23       S(IL,SN)=CC
   21     S(IL,10)=1.
        EE=0.
          DO 26 SN=1,10
          CC=0.
            DO 27 IL=1,9
   27       CC=CC+F(IL)*S(IL,SN)
          Z(IS,SN)=CC-EE
   26     EE=CC
   20   DAV(IS)=DD
   11 CONTINUE
ENDPRO

***** 2. REFLECTION AND EXTINCTION

PROCEDURE EDIFDB,EDIFDV,EDIFDN,KBL,KDFV,KDFN=EXTINC(SUMF,LAID)
STORAGE B(9), RFV(11), RFN(11), KDN(11), KDV(11), KDIR(11)
TABLE B(1-9)=.030,.087,.133,.163,.174,.163,.133,.087,.030
*     DISTRIBUTION OF INCIDENT FLUXES OVER 9 ZONES OF THE SKY (UOC)
      SQVI =SQRT(1.-SCV)
PARAM SCV  =0.2
*     SCATTERING COEFFICIENT OF THE LEAVES IN VISIBLE REGION
      SQNI =SQRT(1.-SCN)
PARAM SCN  =0.85
*     SCATTERING COEFFICIENT OF THE LEAVES IN NEAR-INFRARED REGION
      REFV =(1.-SQVI)/(1.+SQVI)
      REFN =(1.-SQNI)/(1.+SQNI)
        DO 28 IS=1,9
        KDIR(IS+1)=DAV(IS)/SIN( (10*IS-5)*RAD)
        KDN (IS+1)=KDIR(IS+1)*SQNI*0.94623+0.03533
   28   KDV (IS+1)=KDIR(IS+1)*SQVI*0.94623+0.03533
      KDIR( 1)=KDIR( 2)
      KDIR(11)=KDIR(10)
      KDV ( 1)=KDV ( 2)
      KDV (11)=KDV (10)
      KDN ( 1)=KDN ( 2)
      KDN (11)=KDN (10)
```

```
*      DIRECT RADIATION
         DO 29 IS=1,11
         KDIRIS =2.*KDIR(IS)/(KDIR(IS)+1.)
         RFV(IS)=AMAX1(0.,1.117*(1.-EXP(-REFV*KDIRIS) )-0.0111)
   29    RFN(IS)=AMAX1(0.,1.117*(1.-EXP(-REFN*KDIRIS) )-0.0111)

*      DIFFUSE RADIATION
       RFDVV =0.
       RFDVN =0.
       EDIFDB=0.
       EDIFDV=0.
       EDIFDN=0.
         DO 30 J=1,9
         RFDVV =RFDVV +B(J)*RFV(J+1)
         RFDVN =RFDVN +B(J)*RFN(J+1)
         EDIFDB=EDIFDB+B(J)*EXP(-KDIR(J+1)*LAID)
         EDIFDV=EDIFDV+B(J)*EXP(-KDV  (J+1)*LAID)
   30    EDIFDN=EDIFDN+B(J)*EXP(-KDN  (J+1)*LAID)
       KBL    =-ALOG(EDIFDB)/LAID
       KDFV   =-ALOG(EDIFDV)/LAID
       KDFN   =-ALOG(EDIFDN)/LAID
       EDIFIN=EXP(-KDFN*LAI)
       EDIFIV=EXP(-KDFV*LAI)
ENDPRO

***** 3. SITE AND STATE OF CROP

       CSLT =COS(RAD*LAT)
       SNLT =SIN(RAD*LAT)
PARAM LAT  =52.
*      LATITUDE OF SITE

       IWCP =7.88*(IWS+IWR)
*      INITIAL WATER CONTENT OF PLANT, EIGHT TIMES DRY MATTER, TIMES .985
       IWR  =IWOR+IWYR
*      INITIAL WEIGHT OF THE ROOTS
       IWOR =IWS/7.-IWYR
       IWYR =600.*(1.-EXP(-IWS/4200.))
*      IWYR =0.25*IWS
*      IWOR =0.25*IWS
*      ROOT INITIALIZATION FOR YOUNG PLANTS
       IRES =(IWS+IWR)*RESPI/(1.-RESPI)

       IPS=FPS*IWS
       ICS=FCS*IWS
       IFS=FFS*IWS
       ILS=FLS*IWS
       IMS=FMS*IWS
       IAS=FAS*IWS
       IPR=FPR*IWR
       ICR=FCR*IWR
       IFR=FFR*IWR
       ILR=FLR*IWR
       IMR=FMR*IWR
       IAR=FAR*IWR
*      INITIAL AMOUNTS IN SHOOT AND ROOTS,RESPECTIVELY
*      PROTEINS, CARBOHYDRATES, LIPIDS, LIGNIN, MINERALS, ORGANIC ANIONS


       IANCRL=IAMAS*1.629629/86400.
*      INITIAL VALUE OF FIRST ORDER AVERAGE OF CO2 ASSIMILATION
*      RATE (KG CO2/HA/S)
       IAMAS =AMIN1(IWS/3000.,1.)*200.
*      INITIAL AVERAGE METABOLIC ACTIVITY OF THE SHOOT, DEPENDENT
*      ON SHOOT WEIGHT (KG STARCH/HA/DAY)
       IAMAR =IWR/IWS*200.
```

```
*      INITIAL AVERAGE METABOLIC ACTIVITY IN THE ROOT (KG STARCH/HA/DAY)

       LAID =AMIN1(LAI,3.)
       RLAI =LAI-LAID
       RELPRO=1.+2.*AC/AB
*      GEOMETRY FACTOR ACCOUNTING FOR PROJECTION BECAUSE OF SIDE
*      ILLUMINATION BY DIFFUSE LIGHT IN ABSENCE OF DIRECT LIGHT
       DAY  =STDAY
       IVPMO=AFGEN(VPMOTB,BEGIN)
*      INITIAL VALUE OF VPMO
       ITS  =AFGEN(TATB,BEGIN)
*      INITIAL TEMPERATURE OF THE SOIL, EQUAL TO THE AIR TEMPERATURE



DYNAMIC

*************************     SECTION  3     *******************************

*         TIMER VARIABLES

       HOUR =TIME/3600.+AINT(BEGIN)
*      BEGIN ON WHOLE HOURS, IF USING MACRO FOR HOUR TOTALS

*         CLIMATE

***** 1. WEATHER

PROCEDURE SNHSS,TA,TS,DPT,SVP,SLOPE,RH,WS,RA,RR,DRYP=WEATH(DAY,HOUR)

*         DIRECTION OF THE SUN
       DEC  =-23.4*COS(2.*PI*(DAY+10.)/365.)
       SNDC =SIN(RAD*DEC)
       CSDC =COS(RAD*DEC)
       SNHSS=SNLT*SNDC+CSLT*CSDC*COS(PI*(HOUR+11.33)/12.)
*      SINE HEIGHT SUN, TEN DEGREES WEST OF TIME MERIDIAN

*         TEMPERATURE
       TA   =AFGEN(TATB,HOUR)
*      TEMPERATURE OF THE AIR IN DEGREES CENTIGRADE
       TS   =INTGRL(ITS,(TA-TS)/14400.)
*      SOIL TEMPERATURE, FOLLOWS AIR TEMPERATURE WITH DELAY OF 4 HOURS

*         AIR HUMIDITY
       DPT  =AFGEN(DPTTB,HOUR)
       AVP  =6.11*EXP(17.4*DPT/(239.+DPT) )
*      ACTUAL WATER VAPOUR PRESSURE, IN MBAR
       SVP  =6.11*EXP(17.4*TA/(239.+TA) )
*      SATURATION VAPOUR PRESSURE, IN MBAR
       SLOPE=4158.6*SVP/(TA+239.)**2
*      DERIVATIVE OF SATURATION PRESSURE WITH RESPECT TO TEMPERATURE
       RH   =AVP/SVP
*      RELATIVE HUMIDITY

*         WIND SPEED AND RELATED RESISTANCE
       WS   =AFGEN(WSTB,HOUR)
*      WIND SPEED IN THE ENCLOSURE, M/S
       RA   =PARL*SQRT(WDL/WS)*0.5
*      DIFFUSION RESISTANCE OF THE LAMINAR LAYER IN S/M,
*      FACTOR 0.5 ACCOUNTS FOR BOTH SIDES OF THE LEAF
PARAM PARL =185.
       RR   =RA/RHOCP
PARAM RHOCP=1200.
*      HEAT CAPACITY OF THE AIR IN J/M**3 PER KELVIN
       DRYP =(SVP-AVP)/RR
ENDPRO
```

```
***** 2. RADIATION

PROCEDURE DIFOV,DIFCL,SUNDCL,CRC,CRO,FOV,LFOV,LFCL,CRAD,DRAD, ...
          SNHS,PROJ,HSUN,IS=INRAD(LAI,LAID,SNHSS)
      SNHS   =AMAX1(0.,SNHSS)
      HSUN   =ATAN(SNHS/SQRT(1.-SNHS*SNHS) )/RAD
      DIFOV =AFGEN(DFOVTB,HSUN)
      DIFON =0.7*DIFOV
      DIFCL =AFGEN(DFCLTB,HSUN)
      SUNDCL=AFGEN(SUNDTB,HSUN)

FUNCTION DFOVTB=0.,0., 5.,6., 15.,26., 25.,45., 35.,64., 45.,80., ...
                55.,94., 65.,105., 75.,112., 90.,116.
FUNCTION DFCLTB=0.,0., 5.,29., 15.,42., 25.,49., 35.,56., 45.,64., ...
                55.,68., 65.,71., 75.,75., 90.,77.
FUNCTION SUNDTB=0.,0., 5.,0., 15.,88., 25.,175., 35.,262., 45.,336., ...
                55.,402., 65.,452., 75.,483., 90.,504.
*         RADIATION VALUES FOR STANDARD SKIES

      CRC    =(SUNDCL+DIFCL)*2.
      CRO    = DIFOV +DIFON
      CRAD   =AFGEN(CRADTB,HOUR)*700.
*     CRAD   =AFGEN(CRADTB,HOUR)
      DRAD   =INTGRL(0.,CRAD)
*     CUMULATIVE RADIATION, J/M**2
      FOV    =(CRC-CRAD)/(NOT(CRC-CRO)+CRC-CRO)
      LFOV   =LIMIT(0.,1.,FOV)
      LFCL   =1.-LFOV
      IF (SNHS.EQ.0.) GO TO 120

*     DISTRIBUTION
      IS     =(HSUN+10.)/10.
      FISUN  =(HSUN+15.)/10.
      ISUN   =FISUN
      FI     =FISUN-ISUN
      KDR    =KDIR(ISUN)*(1.-FI)+FI*KDIR(ISUN+1)
      KV     =KDV (ISUN)*(1.-FI)+FI*KDV (ISUN+1)
      KN     =KDN (ISUN)*(1.-FI)+FI*KDN (ISUN+1)
      RV     =RFV (ISUN)*(1.-FI)+FI*RFV (ISUN+1)
      RN     =RFN (ISUN)*(1.-FI)+FI*RFN (ISUN+1)
      CSHS   =SQRT(1.-SNHS*SNHS)
      PROJ   =SNHS+AC*CSHS/AB
      ERID   =EXP(-KDR*LAID*SNHS/PROJ)
      EFRIDV=EXP(-KV *LAID*SNHS/PROJ)
      EFRIDN=EXP(-KN *LAID*SNHS/PROJ)
      ERI    =EXP(-KDR*LAI *SNHS/PROJ)
      EFRIV =EXP(-KV *LAI *SNHS/PROJ)
      EFRIN =EXP(-KN *LAI *SNHS/PROJ)
*     EXTINCTION OF ALL TYPES ASSUMED EXPONENTIAL
  120 CONTINUE
ENDPRO


*************************   SECTION  4   ******************************

*         ENERGY BALANCE

PROCEDURE TEHL,TSHL,AVTCP,NCRL,NCRLO,NCRLC,AVIS,ANIR,NETR= ...
          ENERGY(SRW,SNHS,AMAX)

      TEHLO =0.
      TEHLC =0.
      TSHLO =0.
      TSHLC =0.
      AVTCPO=0.
      AVTCPC=0.
```

```
      NCRLO =0.
      NCRLC =0.
      AVISO =0.
      AVISC =0.
      ANIRO =0.
      ANIRC =0.
      IF (CRAD.NE.0.) GO TO 108
      TEHL  =DRYP*LAI/(PSCH*(0.93*RA+RESCW)/RA+SLOPE)
      TSHL  =-TEHL
      AVTCP =TA+TSHL*RR/LAI
      NCRL  =-DPL*LAI/3600.
      GO TO 103
  108 IF (SNHS.NE.0.) GO TO 109
  101 SUNPER=0.
      LFCL  =0.
      LFOV  =1.
      VISDFO=0.8*(1.-RFOVV)*CRAD*(1.-EDIFDV)*RELPRO/(LAID*1.7)
      NIRDFO=0.8*(1.-RFOVN)*CRAD*(1.-EDIFDN)*RELPRO/(LAID*1.7)*0.7
      GO TO 102
  109 SLLA  =AMIN1(PROJ*(1.-ERI)/(SNHS*KDR),LAID)
      SHLA  =LAID-SLLA
      CRADF =FCNSW(LFOV-FOV,CRAD/CRO,1.,CRAD/CRC)*0.8
      PAR   =CRADF*RELPRO*DIFOV/LAID
      VISDFO=(1.-RFOVV)*(1.-EDIFDV)*PAR
      NIRDFO=(1.-RFOVN)*(1.-EDIFDN)*PAR*0.7
      VIST  =SUNDCL*(1.-RV )*(1.-EFRIDV)
      NIRT  =SUNDCL*(1.-RN )*(1.-EFRIDN)
      VISD  =SUNDCL*(1.-SCV)*(1.-ERID  )
      NIRD  =SUNDCL*(1.-SCN)*(1.-ERID  )
      VISDFC=VISDFO*DIFCL/DIFOV+(VIST-VISD)*CRADF*PROJ/(SNHS*LAID)
      NIRDFC=NIRDFO*DIFCL/DIFON+(NIRT-NIRD)*CRADF*PROJ/(SNHS*LAID)
      SUNPER=SUNDCL/SNHS
  102 CONTINUE

*     OVERCAST SKY
      TEHLO,TSHLO,AVTCPO,NCRLO,AVISO,ANIRO=TRPH(VISDFO,NIRDFO,LAID)
      IF (RLAI.EQ.0.) GO TO 104
      VISDFO=VISDFO*(EDIFDV-EDIFIV)*LAID/(RLAI*(1.-EDIFDV) )
      NIRDFO=NIRDFO*(EDIFDN-EDIFIN)*LAID/(RLAI*(1.-EDIFDN) )
      TEHLO,TSHLO,AVTCPO,NCRLO,AVISO,ANIRO=TRPH(VISDFO,NIRDFO,RLAI)

*     CLEAR SKY
  104 IF (LFOV.EQ.1.) GO TO 107
      TEHLC,TSHLC,AVTCPC,NCRLC,AVISC,ANIRC=TRPH(VISDFC,NIRDFC,SHLA)
        DO 106 SN=1,10
        IF (SUMF.NE.0.) ZISSN=Z(IS,SN)
        AREA=SLLA*ZISSN
        BLM =(0.1*SN-0.05)*SUNPER*CRADF
        VIS =VISDFC+BLM*(1.-SCV)
        NIR =NIRDFC+BLM*(1.-SCN)
        TEHLC,TSHLC,AVTCPC,NCRLC,AVISC,ANIRC=TRPH(VIS,NIR,AREA)
  106   CONTINUE

  105 IF (RLAI.EQ.0.) GO TO 107
      VIST  =SUNDCL*(1.-RV )*(EFRIDV-EFRIV)
      NIRT  =SUNDCL*(1.-RN )*(EFRIDN-EFRIN)
      VISD  =SUNDCL*(1.-SCV)*(ERID  -ERI  )
      NIRD  =SUNDCL*(1.-SCN)*(ERID  -ERI  )
      VISDFC=VISDFO*DIFCL/DIFOV+(VIST-VISD)*CRADF*PROJ/(SNHS*RLAI)
      NIRDFC=NIRDFO*DIFCL/DIFON+(NIRT-NIRD)*CRADF*PROJ/(SNHS*RLAI)
      TEHLC,TSHLC,AVTCPC,NCRLC,AVISC,ANIRC=TRPH(VISDFC,NIRDFC,RLAI)
  107 TEHL  = LFOV*TEHLO +LFCL*TEHLC
      TSHL  = LFOV*TSHLO +LFCL*TSHLC
      AVTCP =(LFOV*AVTCPO+LFCL*AVTCPC)/LAI
      NCRL  =(LFOV*NCRLO +LFCL*NCRLC)/3600.
  103 AVIS  = LFOV*AVISO +LFCL*AVISC
```

```
      ANIR  = LFOV*ANIRO +LFCL*ANIRC
      NETR  =AVIS+ANIR
ENDPRO

*************************     SECTION  5    ********************************

*          WATER BALANCE

      RWCP =0.125*WCPL/(WSC+WRC)
*     REL. WATER CONTENT OF CANOPY
      WCPL =INTGRL(IWCP,(WUR-TRC)*10.)
*     WATER CONTENT PLANT IN KG WATER/HA


*************************     SECTION  6    ********************************

*          CROP WATER STATUS

      SRW  =1./AFGEN(SRCTB,RWCP)
      TRCMH=AFGEN(TRCMTB,HOUR)*FLAG
*     TRANSPIRATION RATE OF CANOPY MEASURED (G WATER/M**2/H)
FUNCTION TRCMTB=0.,0., 100.,0.
*     MEASURED TRANSPIRATION RATE TABLE
      FLAG =IMPULS(0.,3600.)*KEEP
      TRCCH=(TTRCC-ZHOLD(IMPULS(DELT,3600.)*KEEP,TTRCC) )*FLAG
*     TRANSPIRATION RATE OF CANOPY CALCULATED, G WATER/M**2/H
      TTRCC=INTGRL(0.,TRC)
*     TOTAL TRANSPIRATION OF CANOPY (G WATER/M**2)
      TRC  =TEHL/VAPHT
*     TRANSPIRATION RATE OF CANOPY IN G/M**2/S
CONST VAPHT=2390.

      TWUR =INTGRL(0.,WUR)
*     WUR  =(WPTSL-WPTC)/(WRESPL+1./ACRS)
      WUR  =(WPTSL-WPTC)*ACRS
*     WATER UPTAKE BY THE ROOTS IN G/M**2/S
PARAM WPTSL=-0.1
*     WATER POTENTIAL OF THE SOIL, 0.1 CORRESPONDS WITH FIELD CAPACITY
      WPTC =AFGEN(WPTTB,RWCP)
*     WATER POTENTIAL CROP IN BAR
      ACRS =(WYR+0.3*WOR)*AFGEN(ETRCTB,TS)/WCRR
*     ACTUAL CONDUCTANCE OF ROOT SYSTEM, G WATER/M**2/BAR/S
PARAM WCRR =2500.
*     WEIGHT/CONDUCTANCE RATIO OF ROOT SYSTEM


*************************     SECTION  7    ********************************


***** 1. PHOTOSYNTHESIS

PARAM EFF=0.50
*     EFFICIENCY AT LOW LIGHT INTENSITIES, KG CO2/HA/H PER J/M**2/S.
      RMES =(RCO2IM-CO2C)*68.4/(AMAX1(0.001,AM300) *REDFRL)
*     AMAX1 PREVENTS DIVIDE CHECK
      AM300=AFGEN(AMTB,TA)
      REDFRL=AFGEN(REDFRT,RESL)
*     REDUCTION FACTOR ACCOUNTING FOR FEEDBACK OF RESERVE LEVEL
*     TO PHOTOSYNTHESIS
FUNCTION REDFRT=0.,1., .20,1., .25,.0001, 1.,.0001
      AMAX =(RCO2I-CO2C)*68.4/RMES
      RCO2I=AMIN1(RCO2IM,RIECO2*VPMOC)
      DPL  =0.11*ANCRL/LAI*3600.
*     DISSIMILATION IN PHOTOSYNTHESIZING LEAVES (KG CO2/HA LEAF/S)
*     USED IN STOMATA REGULATION
      ANCRL=INTGRL(IANCRL,(AMAX1(0.,NCRL)-ANCRL)/(43200.))
```

```
*     FIRST ORDER AVERAGE OF NET CO2-ASSIMILATION

      NCASC=NCAS*3600.
*     NET CO2 ASSIMILATION SHOOT CALCULATED IN KG CO2/HA/H
      NCAS =NCRL-GRS-MRS
*     NET CO2-ASSIMILATION RATE SHOOT IN KG CO2/HA/S

***** 2. RESERVES

      RESL =RES/(RES+WSC+WRC)
*     RESERVE LEVEL
      RES  =INTGRL(IRES,NCRL/1.629629-URES)
*     AMOUNT OF RESERVES (KG STARCH/HA).
      URES =SR+SS+SRSDA+SRDDA
*     USE OF RESERVES


***** 3. EFFICIENCY OF GROWTH

*        SHOOT

      SS     =SRGS+SRMS+SRGSAS+SRTPAS
*     STARCH REQUIREMENT OF THE SHOOT (KG STARCH/HA/S)
      SRGS   =SRGPS+SRGCS+SRGFS+SRGMS+SRGLS
*     STARCH REQUIREMENT FOR GROWTH OF THE SHOOT (KG STARCH/HA/S)
      SRGPS =GRPS*0.517909
      SRGCS =GRCS*1.12
      SRGFS =GRFS*2.73
      SRGLS =GRLS*1.94
      SRGMS =GRMS*0.0184
*     STARCH REQUIREMENT FOR CONVERSION AND TRANSLOCATION OF
*     PROTEINS,CARBOHYDRATES,FATS,LIGNIN AND MINERALS IN THE
*     SHOOT (KG STARCH/HA/S)
      SRGSAS=GRPS*1.25
*     STARCH REQUIREMENT FOR FORMATION OF SKELETONS OF AMINO ACIDS
*     FOR SHOOT PROTEINS (KG STARCH/HA/S)
      SRTPAS=GS*FAS*0.0391
*     STARCH REQUIREMENT FOR TRANSPORT OF ORGANIC ANIONS IN THE
*     SHOOT (KG STARCH/HA/S)

*        ROOT

      SRGPR =GRPR*0.517909
      SRGCR =GRCR*1.12
      SRGFR =GRFR*2.73
      SRGLR =GRLR*1.94
      SRGMR =GRMR*0.0184
*     STARCH REQUIREMENT FOR CONVERSION AND TRANSLOCATION OF
*     PROTEINS,CARBOHYDRATES,FATS,LIGNIN AND MINERALS IN THE
*     ROOT (KG STARCH/HA/S)
      SR     =SRGR+SRMR+SRGSAR+SMU+SRTPAR
*     TOTAL STARCH REQUIREMENT OF THE ROOT (KG STARCH/HA/S)
      SRGR   =SRGPR+SRGCR+SRGFR+SRGLR+SRGMR
*     STARCH REQUIREMENT FOR GROWTH OF THE ROOT
      SRGSAR=GRPR*1.25
*     STARCH REQUIREMENT FOR THE FORMATION OF SKELETONS OF
*     AMINO ACIDS FOR ROOT PROTEINS (KG STARCH/HA/S)
      SRTPAR=GRAR*.0391
*     STARCH REQUIREMENT FOR TRANSPORT OF ORGANIC ANIONS IN THE ROOT
*     (KG STARCH/HA/S)
      SMU    =0.035*(GRMS+GRMR+RNO3)
*     STARCH REQUIREMENT FOR THE UPTAKE OF MINERALS AND NITRATE
*     (KG STARCH/HA/S)

      SRSDA =(RAF-TRRT)*0.8916
```

```
*       STARCH REQUIREMENT FOR FORMATION OF SKELETONS OF ORGANIC ANIONS
        SRDOA =TRRT*0.366341
*       STARCH LOST BY CO2 EVOLUTION DURING DECARBOXYLATION
*       OF ORGANIC ANIONS


***** 4. MAINTENANCE

        SRMS  =((PS*0.0225+MS*0.03)*FNATS+AMAS*0.04)*TEFS/86400.
*       STARCH REQUIREMENT FOR MAINTENANCE OF THE SHOOT (KG STARCH/HA/S)
        FNATS =1.-AMIN1(1.,750.*LAI/WSC)
        AMAS  =INTGRL(IAMAS,(SS-AMAS/86400.)*2.)
*       AVERAGE METABOLIC ACTIVITY SHOOT (KG STARCH PRODUCED AND USED/DAY)
        TEFS  =Q10**(0.1*AVTCP-2.5)
PARAM Q10= 2.
        SRMR  =(PR*0.0225+MR*0.03+AMAR*0.04)*TEFR/86400.
*       STARCH REQUIREMENT FOR MAINTENANCE OF THE ROOT (KG STARCH/HA/S)
        AMAR  =INTGRL(IAMAR,(SR-AMAR/86400.)*2.)
        TEFR  =Q10**(0.1*TS-2.5)

***** 5. CO2 EVOLUTION

        GRS   =GRPS*0.844+GRCS*0.175+GRFS*1.618+GRLS*0.62+GRMS*0.03...
               +SRTPAS*1.629629
*       CO2 EVOLUTION RESULTING FROM GROWTH OF THE SHOOT (KG CO2/HA/S)
        MRS   =SRMS*1.629629
*       CO2 EVOLUTION RESULTING FROM MAINTENANCE OF THE SHOOT(KG CO2/HA/S)
        DR    =GRR+MRR+RUPT+DDA
*       TOTAL CO2-EVOLUTION IN THE ROOT (KG CO2/HA/S)
        GRR   =GRPR*0.844+GRCR*0.175+GRFR*1.618+GRLR*0.62+GRMR*0.03...
               +SRTPAR*1.629629
*       CO2 EVOLUTION RESULTING FROM GROWTH OF THE ROOT (KG CO2/HA/S)
        MRR   =SRMR*1.629629
*       CO2 EVOLUTION RESULTING FROM MAINTENANCE OF THE ROOT(KG CO2/HA/S)
        RUPT  =SMU*1.629629
*       CO2 EVOLUTION RESULTING FROM THE UPTAKE OF MINERALS
        DDA   =TRRT*.597
*       CO2 EVOLUTION RESULTING FROM DECARBOXYLATION OF ORGANIC ANIONS

***** 6. CARBON BALANCE

        RDPF  =(WCP-WCF)/(NOT(WCP)+WCP)
        WCP   =((PS-IPS)+(PR-IPR))*.55555+ ...
               ((CS-ICS)+(CR-ICR))*.45005+ ...
               ((FS-IFS)+(FR-IFR))*.77206+ ...
               ((LS-ILS)+(LR-ILR))*.69313+ ...
               ((AS-IAS)+(AR-IAR))*.39627
        WCF   =TNCAP*.272727+(IRES-RES)*.444444
*       WCP AND WCF IN KG CARBON/HA, THE DIFFERENCE BETWEEN BOTH MUST BE
*       SMALLER THAN 0.01 TIMES THEIR VALUE
        TNCAP=INTGRL(0.,NCAS-DR)
*       TOTAL NET CO2-ASSIMILATION PLANT (KG CO2/HA)


*************************    SECTION  8    ********************************


***** 1. CROP GROWTH

        TWT   =WSC+WRC+RES
*       TOTAL DRY WEIGHT (KG/HA)
        TWS   =WSC+RES
*       TOTAL WEIGHT SHOOT
```

```
***** 2. SHOOT

      WSC  =PS+CS+FS+LS+MS+AS
*     WEIGHT SHOOT CALCULATED

      PS   =INTGRL(IPS,GRPS)
      CS   =INTGRL(ICS,GRCS)
      FS   =INTGRL(IFS,GRFS)
      LS   =INTGRL(ILS,GRLS)
      MS   =INTGRL(IMS,GRMS)
      AS   =INTGRL(IAS,GRAS)
*     WEIGHT OF PROTEINS,CARBOHYDRATES,FATS,LIGNIN,MINERALS AND
*     ORGANIC ANIONS IN THE SHOOT (KG/HA)

      GRPS =GS*FPS
      GRCS =GS*FCS
      GRFS =GS*FFS
      GRLS =GS*FLS
      GRMS =GS*FMS
      GRAS =GS*FAS
*     GROWTH RATE OF PROTEINS,CARBOHYDRATES,FATS,LIGNIN,MINERALS AND
*     ORGANIC ANIONS IN THE SHOOT (KG/HA/S)

      GS   =RES*RCRS*AFGEN(TGTB,AVTCP)*AFGEN(WGSTB,RWCP)
*     GROWTH RATE OF THE SHOOT (KG DM/HA/S)
PARAM RCRS=1.3E-5
*     RELATIVE CONSUMPTION RATE OF THE RESERVES

*     TGTB REPRESENTS INFLUENCE OF TEMPERATURE ON GROWTH (SPECIES DEPEND
ENT)
FUNCTION WGSTB=0.,0., .8,0., .9,.15, .975,.5, 1.,1.
*     WGSTB ACCOUNTS FOR CHANGES IN PARTITIONING OF RESERVES BETWEEN
*     SHOOT AND ROOT, UNDER INFLUENCE OF WATER CONTENT

      RAF  =RNO3*1.06
*     RATE OF FORMATION OF ORGANIC ANIONS,CONCURRENT WITH
*     RATE OF NITRATE REDUCTION
      RNO3 =(GS*FPS+GYR*FPR)*0.652
*     RATE OF NITRATE REDUCTION, ASSUMING THAT ALL PROTEIN N ORIGINATES
*     FROM NITRATES (KG NO3/HA/S)
      TRRT =RAF-GS*FAS-GYR*FAR
*     RATE OF TRANSPORT OF ORGANIC ANIONS TO THE ROOT FOR
*     DECARBOXYLATION (KG/HA/S)

***** 3. ROOT

      WRC  =WOR+WYR
*     WEIGHT OF ROOTS CALCULATED
      WOR  =INTGRL(IWOR,SYR-WOR/ROTC)
*     WEIGHT OF OLD ROOTS (SUBERIZED)
      SYR  =WYR*AFGEN(ETRCTB,TS)/SUBC
*     SUBERIZATION RATE OF YOUNG ROOTS (KG/HA/S)
PARAM SUBC =4.50E5
*     TIME CONSTANT FOR SUBERIZATION OF YOUNG ROOTS (= 5 DAYS)
*     ETRCTB ACCOUNTS FOR THE INFLUENCE OF TEMPERATURE ON THE RATE
*     OF SUBERIZATION
PARAM ROTC =30.E6
*     TIME CONSTANT FOR ROTTING OF OLD ROOTS (ABOUT ONE YEAR, SO
*     INEFFECTIVE HERE)
      WYR  =INTGRL(IWYR,GYR-SYR)
*     WEIGHT OF YOUNG ROOTS (KG DM/HA)
      GYR  =RES*RCRS*AFGEN(TGTB,TS)*AFGEN(WGRTB,RWCP)
*     GROWTH RATE OF YOUNG ROOTS (KG DM/HA/S)
FUNCTION WGRTB=0.,1., .8,1., .9,.85, .975,.5, 1.,0.
*     WGRTB ACCOUNTS FOR THE INFLUENCE OF WATER CONTENT ON THE
```

```
*     PARTITIONING BETWEEN SHOOT AND ROOT

      PR    =INTGRL(IPR,GRPR)
      CR    =INTGRL(ICR,GRCR)
      FR    =INTGRL(IFR,GRFR)
      LR    =INTGRL(ILR,GRLR)
      MR    =INTGRL(IMR,GRMR)
      AR    =INTGRL(IAR,GRAR)
*     AMOUNT OF PROTEINS,CARBOHYDRATES,FATS,LIGNIN,MINERALS AND
*     ORGANIC ANIONS IN THE ROOT (KG DM/HA)

      GRPR  =GYR*FPR
      GRCR  =GYR*FCR
      GRFR  =GYR*FFR
      GRLR  =GYR*FLR
      GRMR  =GYR*FMR
      GRAR  =GYR*FAR
*     GROWTH RATE OF PROTEINS,CARBOHYDRATES,FATS,LIGNIN,MINERALS AND
*     ORGANIC ANIONS IN THE ROOT (KG DM/HA/S)


************************      SECTION  9      ****************************

***** 1. DATA FROM ASSIMILATION CHAMBER

      VPMOC=INTGRL(300.,LTAIR*(VPMI-VPMOC)*0.001-NCAS*BAKOPP*54.54545)
*     OVERALL CO2 CONCENTRATION CALCULATED IN VPPM
      LTAIR=AFGEN(LTAITB,HOUR)/3600.
*     LITRES OF AIR SUPPLIED PER SECOND
      VPMI =AFGEN(VPMITB,HOUR)
      NCASM=0.066*LTAIR*(VPMI-VPMOM)/BAKOPP
*     NET CO2-ASSIMILATION SHOOT MEASURED IN KG CO2/HA/H
      VPMOM=AFGEN(VPMOTB,HOUR)

*     RELPRO=1.+2.*AC/AB
*     GEOMETRY FACTOR ACCOUNTING FOR PROJECTION BECAUSE OF SIDE
*     ILLUMINATION BY DIFFUSE LIGHT IN ABSENCE OF DIRECT LIGHT

***** 2. OUTPUT

NOSORT
CALL DEBUG(2,0.)
      IF (KEEP.LT.0.5) GO TO 200
      TTRCM=TTRCM+TRCMH
      IF (TIME.EQ.0.) TTRCM=0.
  200 TELLER=TELLER+1.
      IF (TIME.EQ.0.) TELLER=0.

PRINT    DAY  ,    DRAD ,    NCRL ,    RWCP ,    ...
         HOUR ,    CRAD ,    AMAS ,    SRW  ,    ...
         TRC  ,    ANIR ,    NCAS ,    RA   ,    ...
         WUR  ,    AVIS ,    ANCRL,    ACRS ,    ...
         TEHL ,    NETR ,    NCRLD,    WSC  ,    ...
         TSHL ,    TA   ,    NCRLC,    WRC  ,    ...
         TTRCC,    FOV  ,    DPL  ,    WYR  ,    ...
         TTRCM,    RDPF ,    GS   ,    RES  ,    ...
         AVTCP,    SNHS ,    DR   ,    VPMOC,    ...
         VPMOM,    VPMI ,    NCASC,    NCASM,    ...
         TRCCH,    TRCMH,    RCO2I
OUTPUT NCASM,NCASC,CRAD
PAGE GROUP=2

***** 3. RUN CONTROL

FINISH TELLER=6000.,AVTCP=-1.,AVTCP=100.,NCAS=-100.,RDPF=0.01
```

```
TIMER FINTIM =86400.,PRDEL =1800.,OUTDEL =900.
◆     TIME IS EXPRESSED IN SECONDS
METHOD RKS

◆◆◆◆◆ 4. EXCHANGEABLE DATA

FUNCTION LTAITB = 11.000, 9.2477E3 ...
,11.167, 9.2477E3,11.500,13.8793E3,11.667,13.8793E3,11.883,21.0514E3 ...
,12.050,21.0514E3,12.233,28.7219E3,12.400,28.7219E3,12.500,28.7219E3 ...
,12.667,28.7219E3,13.083,28.7219E3,13.250,28.7219E3,13.333,28.7219E3 ...
,13.500,28.7219E3,13.833,28.7219E3,14.000,28.7219E3,14.500,28.7219E3 ...
,14.667,28.7219E3,15.333,28.7219E3,15.500,28.7219E3,16.333,28.7219E3 ...
,16.500,28.7219E3,17.333,28.7219E3,17.500,28.7219E3,18.333,28.7219E3 ...
,18.500,28.7219E3,20.333,28.9809E3,20.500,28.9809E3,21.833,28.9809E3 ...
,22.000,28.9809E3,23.833,28.9809E3,24.000,28.9809E3,25.833,28.9809E3 ...
,26.000,28.9809E3,27.833,28.9809E3,28.000,28.9809E3,30.333,28.9809E3 ...
,30.500,28.9809E3,31.833,28.9809E3,32.000,28.9809E3,32.333,28.9809E3 ...
,32.500,28.9809E3
FUNCTION CRADTB = 11.000, 0.9834E0 ...
,11.167, 0.9834E0,11.500, 1.0132E0,11.667, 1.0132E0,11.883, 1.0579E0 ...
,12.050, 1.0579E0,12.233, 1.0802E0,12.400, 1.0802E0,12.500, 1.0802E0 ...
,12.667, 1.0802E0,13.083, 1.0802E0,13.250, 1.0802E0,13.333, 1.0802E0 ...
,13.500, 1.0802E0,13.833, 1.0653E0,14.000, 1.0653E0,14.500, 1.0058E0 ...
,14.667, 1.0058E0,15.333, 0.8791E0,15.500, 0.8791E0,16.333, 0.6929E0 ...
,16.500, 0.6929E0,17.333, 0.4619E0,17.500, 0.4619E0,18.333, 0.2309E0 ...
,18.500, 0.2309E0,20.333, 0.0000E0,20.500, 0.0000E0,21.833, 0.0000E0 ...
,22.000, 0.0000E0,23.833, 0.0000E0,24.000, 0.0000E0,25.833, 0.0000E0 ...
,26.000, 0.0000E0,27.833, 0.0000E0,28.000, 0.0000E0,30.333, 0.0820E0 ...
,30.500, 0.0820E0,31.833, 0.3800E0,32.000, 0.3800E0,32.333, 0.4917E0 ...
,32.500, 0.4917E0
FUNCTION TATB   = 11.000, 2.4375E1 ...
,11.167, 2.4375E1,11.500, 2.5313E1,11.667, 2.5313E1,11.883, 2.6250E1 ...
,12.050, 2.6250E1,12.233, 2.6250E1,12.400, 2.6250E1,12.500, 2.6563E1 ...
,12.667, 2.6563E1,13.083, 2.7813E1,13.250, 2.7813E1,13.333, 2.7188E1 ...
,13.500, 2.7188E1,13.833, 2.7813E1,14.000, 2.7813E1,14.500, 2.7813E1 ...
,14.667, 2.7813E1,15.333, 2.8125E1,15.500, 2.8125E1,16.333, 2.7188E1 ...
,16.500, 2.7188E1,17.333, 2.6875E1,17.500, 2.6875E1,18.333, 2.5625E1 ...
,18.500, 2.5625E1,20.333, 2.1250E1,20.500, 2.1250E1,21.833, 2.0000E1 ...
,22.000, 2.0000E1,23.833, 2.0000E1,24.000, 2.0000E1,25.833, 1.8062E1 ...
,26.000, 1.8062E1,27.833, 1.6937E1,28.000, 1.6937E1,30.333, 1.3750E1 ...
,30.500, 1.3750E1,31.833, 1.6563E1,32.000, 1.6563E1,32.333, 1.8750E1 ...
,32.500, 1.8750E1
FUNCTION VPMOTB = 11.000, 1.1819E2 ...
,11.167, 1.1819E2,11.500, 1.6139E2,11.667, 1.6139E2,11.883, 2.0126E2 ...
,12.050, 2.0126E2,12.233, 2.2710E2,12.400, 2.2710E2,12.500, 2.6955E2 ...
,12.667, 2.6955E2,13.083, 3.2419E2,13.250, 3.2419E2,13.333, 3.1746E2 ...
,13.500, 3.1746E2,13.833, 3.8660E2,14.000, 3.8660E2,14.500, 4.3301E2 ...
,14.667, 4.3301E2,15.333, 4.3396E2,15.500, 4.3396E2,16.333, 4.5053E2 ...
,16.500, 4.5053E2,17.333, 4.7421E2,17.500, 4.7421E2,18.333, 5.1115E2 ...
,18.500, 5.1115E2,20.333, 3.2930E2,20.500, 3.2930E2,21.833, 3.2788E2 ...
,22.000, 3.2788E2,23.833, 3.3545E2,24.000, 3.3545E2,25.833, 3.3924E2 ...
,26.000, 3.3924E2,27.833, 3.4019E2,28.000, 3.4019E2,30.333, 3.3640E2 ...
,30.500, 3.3640E2,31.833, 5.1351E2,32.000, 5.1351E2,32.333, 5.0262E2 ...
,32.500, 5.0262E2
FUNCTION VPMITB = 11.000, 3.1201E2 ...
,11.167, 3.1201E2,11.500, 3.1016E2,11.667, 3.1016E2,11.883, 3.0831E2 ...
,12.050, 3.0831E2,12.233, 3.0721E2,12.400, 3.0721E2,12.500, 3.5261E2 ...
,12.667, 3.5261E2,13.083, 4.0762E2,13.250, 4.0762E2,13.333, 4.0744E2 ...
,13.500, 4.0744E2,13.833, 4.7421E2,14.000, 4.7421E2,14.500, 5.2346E2 ...
,14.667, 5.2346E2,15.333, 5.2299E2,15.500, 5.2299E2,16.333, 5.2299E2 ...
,16.500, 5.2299E2,17.333, 5.2299E2,17.500, 5.2299E2,18.333, 5.3814E2 ...
,18.500, 5.3814E2,20.333, 3.1699E2,20.500, 3.1699E2,21.833, 3.1888E2 ...
,22.000, 3.1888E2,23.833, 3.2646E2,24.000, 3.2646E2,25.833, 3.3119E2 ...
,26.000, 3.3119E2,27.833, 3.3356E2,28.000, 3.3356E2,30.333, 3.3877E2 ...
,30.500, 3.3877E2,31.833, 5.4856E2,32.000, 5.4856E2,32.333, 5.4524E2 ...
,32.500, 5.4524E2
```

```
FUNCTION DPTTB=      11.16,16.50 ,  11.66,16.00 ,  12.05,15.00 ,...
      12.40,14.00 ,  12.66,13.50 ,  13.25,13.50 ,  13.50,13.00 ,...
      14.00,13.00 ,  14.66,13.50 ,  15.50,14.00 ,  16.50,12.00 ,...
      17.50,16.50 ,  18.50,14.00 ,  20.50,12.00 ,  22.00,10.00 ,...
      24.00,10.00 ,  26.00, 8.50 ,  28.00, 8.00 ,  30.50, 6.50 ,...
      32.00, 7.00 ,  32.50, 8.00
FUNCTION TRCMTB = 11.,0.,12.,604.9,13.,740.4,14.,581.0,15.,520.8,    ...
      16.,498.1,17.,460.4,18.,454.5,19.,351.2,20.,218.5,21.,122.0,    ...
      22.,67.2,23.,96.6,24.,96.6,25.,86.6,26.,77.1,27.,68.7,28.,69.2, ...
      29.,25.8,30.,38.4,31.,40.1,32.,91.8
*     EXPERIMENT PRN 346
END
STOP
ENDJOB
```

Appendix C – List of abbreviations

NAME	DESCRIPTION	UNIT
AA	INTERMEDIATE VARIABLE	-
AB	SIDE ENCLOSURE	M
ABSRAD	TOTAL ABSORBED RADIATIVE ENERGY PER LEAF AREA	J/M2/S
AC	HEIGHT CANOPY	M
ACRS	ACTUAL CONDUCTANCE OF THE ROOT SYSTEM	G(H2O)/M2/BAR/S
ALWR	LONG WAVE OR THERMAL RADIATION ABSORBED BY CANOPY	J/M2/S
ALWRO	THERMAL RADIATION OVERCAST SKY ABSORBED BY CANOPY	J/M2/S
ALWRC	THERMAL RADIATION CLEAR SKY ABSORBED BY CANOPY	J/M2/S
AM300	MAXIMUM PHOTOSYNTHESIS AT 300 VPPM	KG(CO2)/HA(LEAF)/
AMAR	AVERAGE METABOLIC ACTIVITY ROOT	KG(STARCH)/HA/DA
AMAS	AVERAGE METABOLIC ACTIVITY SHOOT	KG(STARCH)/HA/DA
AMAX	CO2 ASSIMILATION RATE OF A LEAF AT LIGHT SATURATION	KG(CO2)/HA(LEAF)/
AMAX1	CSMP FUNCTION TAKES THE GREATEST OF THE TWO ARGUMENTS	-
AMTB	TABLE OF AMAX VERSUS AIR TEMPERATURE AT 300 VPPM CO2	-
ANCRL	FIRST ORDER AVERAGE OF NET CO2-ASSIMILATION	KG/HA/S
ANETR	NET RADIATION ABSORBED BY CANOPY	J/M2/S
ANETRC	NET RADIATION ABSORBED BY CANOPY UNDER CLEAR SKY	J/M2/S
ANETRO	NET RADIATION ABSORBED BY CANOPY UNDER OVERCAST SKY	J/M2/S
ANIR	NEAR-INFRARED RADIATION,ABSORBED BY CANOPY,PER GROUND AREA	J/M2/S
ANIRC	ABSORBED NEAR-INFRARED UNDER A CLEAR SKY	J/M2/S
ANIRO	ABSORBED NEAR-INFRARED UNDER AN OVERCAST SKY	J/M2/S
AR	AMOUNT OF ORGANIC ANIONS IN ROOTS	KG/HA
AREA	AREA OF LEAVES IN A CERTAIN IRRADIATION CLASS	M2(LEAF)/M2(GROUND
AS	AMOUNT OF ORGANIC ANIONS IN SHOOT	KG/HA
AVIS	VISIBLE RADIATION,ABSORBED BY CANOPY,PER GROUND AREA	J/M2/S
AVISC	ABSORBED VISIBLE RADIATION UNDER A CLEAR SKY	J/M2/S
AVISO	ABSORBED VISIBLE RADIATION UNDER AN OVERCAST SKY	J/M2/S
AVP	ACTUAL WATER VAPOUR PRESSURE	MBAR
AVTCP	AVERAGE TEMPERATURE OF ALL LEAVES	DEGREE(C)
AVTCPC	AVERAGE TEMPERATURE CANOPY UNDER CLEAR SKY	DEGREE(C)
AVTCPO	AVERAGE TEMPERATURE CANOPY UNDER OVERCAST SKY	DEGREE(C)
B	RELATIVE CONTRIBUTION OF NINE ZONES OF UNIFORM OVERCAST SKY (ISOTROPIC)	-
BAKOPP	GROUND AREA OF ENCLOSURE	M2
BB	INTERMEDIATE VARIABLE	-
BEGIN	HOUR WHEN SIMULATION BEGINS	H
BLM	INTERMEDIATE NAME FOR SINE OF INCIDENCE OF SUNLIGHT ON LEAF	-
BOWRAT	BOWEN'S RATIO	-
CO	INTERMEDIATE NAME FOR COSINE OF LEAF INCLINATION	-
CO2C	CO2 COMPENSATION POINT	VPPM
CR	AMOUNT OF CARBOHYDRATES IN ROOTS	KG/HA
CRAD	CURRENT GLOBAL RADIATION	J/M2/S
CRADF	CURRENT GLOBAL RADIATION RELATED TO CRC AND CRO	-
CRADTB	TABLE OF MEASURED GLOBAL RADIATION AS FUNCTION OF TIME	-
CRC	CURRENT GLOBAL RADIATION CLEAR	J/M2/S
CRO	CURRENT GLOBAL RADIATION OVERCAST	J/M2/S
CS	AMOUNT OF CARBOHYDRATES IN SHOOT	KG/HA
CSDC	COSINE OF DECLINATION	-

NAME	DESCRIPTION	UNIT
CSHS	COSINE OF SUN HEIGHT	-
CSLT	COSINE OF LATITUDE	-
DAY	NUMBER OF DAY IN THE YEAR FROM 1ST OF JANUARY	DAY
DDA	CO2 EVOLUTION RESULTING FROM DECARBOXYLATION OF ORGANIC ANIONS	KG(CO2)/HA/S
DEC	DECLINATION OF SUN WITH RESPECT TO THE EQUATOR	DEGREE
DELT	TIME STEP OF INTEGRATION	S
DFCLTB	DIFFUSE VISIBLE RADIATION STANDARD SKY CLEAR TABLE	-
DFOVTB	DIFFUSE VISIBLE RADIATION STANDARD SKY OVERCAST TABLE	-
DIFCL	DIFFUSE VISIBLE RADIATION STANDARD SKY CLEAR	J/M2/S
DIFF	DIFFERENCE BETWEEN TRANSPIRATION OF CANOPY	G(H2O)/HA/S
DIFF1	DIFFERENCE BETWEEN TRANSPIRATION OF CANOPY AND WATER UPTAKE RATE	G(H2O)/HA/S
DIFON	DIFFUSE NEAR-INFRARED RADIATION STANDARD SKY OVERCAST	J/M2/S
DIFOV	DIFFUSE VISIBLE RADIATION STANDARD SKY OVERCAST	J/M2/S
DLYTOT	NAME OF MACRO TO CALCULATE DAILY TOTAL	-
DNETRS	DAILY TOTAL OF NETRS	J/M2/DAY
DPL	DISSIMILATION RATE OF LEAVES THAT PHOTOSYNTHESIZE IN DAYTIME	KG(CO2)/HA(LEAF)/S
DPT	DEW POINT TEMPERATURE	DEGREE(C)
DPTC	DEW POINT TEMPERATURE	DEGREE(C)
DPTTB	TABLE OF DEW POINT TEMPERATUREAS A FUNCTION OF TIME	-
DR	TOTAL CO2-EVOLUTION OF THE ROOT	KG(CO2)/HA/S
DRC	DAILY TOTAL GLOBAL RADIATION CLEAR	J/M2
DRCI	INITIAL VALUE OF DRC	J/M2
DRCP	DRC ON PREVIOUS DAY	J/M2
DRAD	ACCUMULATED GLOBAL RADIATION	J/M2
DRO	DAILY TOTAL GLOBAL RADIATION OVERCAST	J/M2
DROI	INITIAL VALUE OF DRO	J/M2
DROP	DRO ON PREVIOUS DAY	J/M2
DRYP	DRYING POWER OF AIR	MBAR J/M2/S/DEGREE(C)
DTABR	DAILY TOTAL RADIATION ABSORBED BY CANOPY	J/M2/DAY
DTGSC	DAILY TOTAL GROWTH SHOOT CALCULATED	KG/HA/DAY
DTGSM	DAILY TOTAL GROWTH SHOOT MEASURED	KG/HA/DAY
DTLWR	DAILY TOTAL LONG WAVE RADIATION	J/M2/DAY
DTOTI	INTERMEDIATE VARIABLE IN MACRO TO CALCULATE DAILY TOTAL	-
DTOT	INTERMEDIATE VARIABLE IN MACRO TO CALCULATE DAILY TOTAL	-
DTOT1	INTERMEDIATE VARIABLE IN MACRO TO CALCULATE DAILY TOTAL	-
DTR	DAILY TOTAL GLOBAL RADIATION MEASURED	J/M2/DAY
DTRT	DAILY TOTAL MEASURED GLOBAL RADIATION FUNCTION VERSUS TIME (OTHER DIMENSIONS LIKE CAL/CM2 OR COUNTS ARE ALSO POSSIBLE)	J/M2/DAY
DTTRC	CALCULATED DAILY TOTAL OF TRANSPIRATION	MM OR KG(H2O)/M2/DAY
ECO2C	EXTERNAL CO2-CONCENTRATION	VPPM
EDIFB	TRANSMITTED DIFFUSE BLACK INFRARED	-
EDIFDB	TRANSMITTED -BLACK-DIFFUSE RADIATION	-
EDIFDN	TRANSMITTED DIFFUSE NEAR-INFRARED	-
EDIFDV	TRANSMITTED DIFFUSE VISIBLE RADIATION	-
EDIFIN	TRANSMITTED DIFFUSE NEAR-INFRARED	-
EDIFIV	TRANSMITTED DIFFUSE VISIBLE RADIATION	-
EDIFN	TRANSMITTED DIFFUSE NEAR-INFRARED	-
EDIFV	TRANSMITTED DIFFUSE VISIBLE RADIATION	-
EDPTTB	DEW POINT TEMPERATUREAT SUN RISE VERSUS TIME	-
EFF	EFFICIENCY OF CO2 ASSIMILATION DERIVATIVE OF CO2 ASSIMILATION VERSUS ABSORBED VISIBLE RADIATION AT LOW LIGHT INTENSITY	KG(CO2)/J/HA/H M2S
EFRIDN	TRANSMITTED NEAR-INFRARED RADIATION (DIRECT)	-
EFRIDV	TRANSMITTED VISIBLE RADIATION (DIRECT)	-
EFRIN	TRANSMITTED NEAR-INFRARED RADIATION (DIRECT)	-

NAME	DESCRIPTION	UNIT
EFRIV	TRANSMITTED VISIBLE RADIATION (DIRECT)	-
EHL	EVAPORATIVE HEAT LOSS OF LEAVES PER LEAF AREA	J/M2(LEAF)/
ENP	ENERGY USED FOR CO2 ASSIMILATION PER LEAF AREA	J/M2(LEAF)/
ERI	TRANSMITTED DIRECT RADIATION	-
ERID	TRANSMITTED DIRECT RADIATION	-
ERROR	ACCEPTED DIFFERENCE BETWEEN TRANSPIRATION RATE AND WATER UPTAKE IN ITERATION PROCEDURE	-
ETRCTB	EFFECT OF TS ON CONDUCTANCE OF ROOT SYSTEM TABLE	
EVA	INTERMEDIATE VARIABLE COMPRISING EFFICIENCY OF INCOMING RADIATION AND MAXIMUM CO2-ASSIMILATION	-
F	LEAF ANGLE DISTRIBUTION	-
FAR	FRACTION OF ORGANIC ANIONS IN ROOTS CURRENTLY SYNTHESIZED	-
FAS	FRACTION OF ORGANIC ANIONS IN SHOOTS CURRENTLY SYNTHESIZED	-
FATB	FRACTION OF ORGANIC ANIONS TABLE VERSUS TIME	-
FCL	FRACTION OF OF TIME THAT SKY IS CLEAR	-
FCNSW	CSMP FUNCTION	-
FCR	FRACTION OF STRUCTURAL CARBOHYDRATES IN ROOTS CURRENTLY SYNTHESIZED	-
FCS	FRACTION OF STRUCTURAL CARBOHYDRATES IN SHOOTS CURRENTLY SYNTHESIZED	-
FCTB	FRACTION OF CARBOHYDRATES TABLE VERSUS TIME	-
FFR	FRACTION OF FATS IN ROOTS CURRENTLY SYNTHESIZED	-
FFS	FRACTION OF FATS IN SHOOTS CURRENTLY SYNTHESIZED	-
FFTB	FRACTION OF FATS TABLE VERSUS TIME	-
FGNS	FRACTION OF SHOOT GROWTH OCCURRING IN NON-ASSIMILATING SHOOT PARTS	-
FI	ANGLE OF INCOMING SUN RAYS	DEGREE
FIC	FRACTION OF PLANT CONSTITUENTS IN MACRO INCREMENT	-
FINTIM	TOTAL DURATION OF SIMULATION RUN	S
FISUN	ANGLE OF INCOMING SUN RAYS	DEGREE
FLAG	CONTROL VARIABLE	-
FLR	FRACTION OF LIGNIN IN ROOTS CURRENTLY SYNTHESIZED	-
FLS	FRACTION OF LIGNIN IN SHOOTS CURRENTLY SYNTHESIZED	-
FLTB	FRACTION OF LIGNIN TABLE VERSUS TIME	-
FMR	FRACTION OF MINERALS IN ROOTS CURRENTLY SYNTHESIZED	-
FMS	FRACTION OF MINERALS IN SHOOTS CURRENTLY SYNTHESIZED	-
FMTB	FRACTION OF MINERALS TABLE VERSUS TIME	-
FNATS	FRACTION OF NON-ASSIMILATING TISSUE	-
FOV	FRACTION OF TIME THAT SKY IS OVERCAST	-
FPR	FRACTION OF PROTEINS IN ROOTS CURRENTLY SYNTHESIZED	-
FPS	FRACTION OF PROTEINS IN SHOOTS CURRENTLY SYNTHESIZED	-
FPTB	FRACTION PROTEINS TABLE VERSUS TIME	-
FR	AMOUNT OF FATS IN ROOTS	KG/HA
FS	AMOUNT OF FATS IN SHOOT	KG/HA
FT	FRACTION	-
GRAR	GROWTH RATE OF ORGANIC ANIONS IN THE ROOT	KG/HA/S
GRAS	GROWTH RATE OF ORGANIC ANIONS IN THE SHOOT	KG/HA/S
GRCR	GROWTH RATE OF CARBOHYDRATES IN THE ROOT	KG/HA/S
GRCS	GROWTH RATE OF CARBOHYDRATES IN THE SHOOT	KG/HA/S
GRFR	GROWTH RATE OF FATS IN THE ROOT	KG/HA/S
GRFS	GROWTH RATE OF FATS IN THE SHOOT	KG/HA/S
GRLR	GROWTH RATE OF LIGNIN IN THE ROOT	KG/HA/S
GRLS	GROWTH RATE OF LIGNIN IN THE SHOOT	KG/HA/S
GRMR	GROWTH RATE OF MINERALS IN THE ROOT	KG/HA/S
GRMS	GROWTH RATE OF MINERALS IN THE SHOOT	KG/HA/S
GRPR	GROWTH RATE OF PROTEIN IN THE ROOT	KG/HA/S
GRPS	GROWTH RATE OF PROTEINS IN THE SHOOT	KG/HA/S
GRR	CO2 EVOLUTION RESULTING FROM GROWTH OF THE ROOT	KG/HA/S
GRS	CO2 EVOLUTION RESULTING FROM GROWTH OF THE SHOOT	KG/HA/S

NAME	DESCRIPTION	UNIT
GS	GROWTH SHOOT (UNADJUSTED)	KG/HA/S
GSM	GROWTH SHOOT MEASURED	KG/HA/S
GYR	GROWTH RATE OF YOUNG ROOTS (UNADJUSTED)	KG/HA/S
HOUR	TIME OF THE DAY IN HOURS	H
HSUN	HEIGHT OF THE SUN	DEGREE
I	RUNNER IN DO LOOP	-
IA	INITIAL AMOUNT OF ORGANIC ACIDS IN SHOOT AND ROOT	KG/HA
IAMAR	INITIAL VALUE OF AMAR	KG(STARCH)/HA/DAY
IAMAS	INITIAL VALUE OF AMAS	KG(STARCH)/HA/DAY
IANCRL	INITIAL VALUE OF ANCRL	KG/HA/S
IAR	INITIAL VALUE OF AR	KG/HA
IAS	INITIAL VALUE OF AS	KG/HA
ICR	INITIAL VALUE OF CR	KG/HA
ICS	INITIAL VALUE OF CS	KG/HA
IFR	INITIAL VALUE OF FR	KG/HA
IFS	INITIAL VALUE OF FS	KG/HA
IL	NUMBER OF INCLINATION CLASS OF LEAF	-
ILR	INITIAL VALUE OF LR	KG/HA
ILS	INITIAL VALUE OF LS	KG/HA
IMR	INITIAL VALUE OF MR	KG/HA
IMS	INITIAL VALUE OF MS	KG/HA
IPR	INITIAL VALUE OF PR	KG/HA
IPS	INITIAL VALUE OF PS	KG/HA
IRES	INITIAL OF RES	KG(STARCH)/HA
IS	NUMBER OF INCLINATION CLASS OF SUN	-
ISUN	NUMBER OF INCLINATION CLASS OF SUN,SHIFTED 5 DEGREES	-
ITS	INITIAL VALUE OF TS	DEGREE(C)
IVPMD	INITIAL VALUE OF VPMDC	VPPM
IWCP	INITIAL VALUE OF WCPL	KG(H2O)/HA
IWOR	INITIAL VALUE OF WOR	KG(DM)/HA
IWR	INITIAL VALUE OF WRC	KG(DM)/HA
IWS	INITIAL VALUE OF WSC	KG(DM)/HA
IWYR	INITIAL VALUE OF WYR	KG(DM)/HA
J	RUNNER IN DO LOOP	-
K	EXTINCTION COEFFICIENT (IN GENERAL)	1/LAI
KBL	EXTINCTION COEFFICIENT FOR DIFFUSE RADIATION AND BLACK LEAVES	1/LAI
KDFN	EXTINCTION COEFFICIENT FOR DIFFUSE NEAR-INFRARED	1/LAI
KDFV	EXTINCTION COEFFICIENT FOR DIFFUSE VISIBLE RADIATION	1/LAI
KDIR	EXTINCTION COEFFICIENT FOR DIRECT RADIATION AND BLACK LEAVES	1/LAI
KDIRIS	EXTINCTION COEFFICIENT FOR DIRECT RADIATION IN CANOPY WITH HORIZONTAL LEAVES	1/LAI
KDN	EXTINCTION COEFFICIENT UNDER DIRECT IRRADIATION NEAR-INFRARED	1/LAI
KDR	EXTINCTION COEFFIENT FOR DIFFUSE DIRECT RADIATION	1/LAI
KDV	EXTINCTION COEFFICIENT UNDER DIRECT IRRADIATION VISIBLE	1/LAI
KEEP	INTERNAL CSMP VARIABLE ,1 IF INTEGRATION IS PERFORMED	-
KN	SEE KDN	1/LAI
KV	SEE KDV	1/LAI
L	RUNNER IN DO LOOP	-
LAI	LEAF AREA INDEX	M2(LEAF)/M2(GROUND)
LAIC	LEAF AREA INDEX, RECKONED FROM ABOVE	M2/M2
LAID	THE UPPER PART OF LAI, UP TO A MAXIMUM OF 3	M2/M2
LAIR	DECIMAL PART OF LAI	-
LAITB	LEAF AREA INDEX TABLE VERSUS TIME	-

NAME	DESCRIPTION	UNIT
LAT	LATITUDE OF EXPERIMENTAL PLOT	DEGREE
LR	AMOUNT OF LIGNIN IN ROOTS	KG/HA
LFCL	COMPLEMENT OF LFOV	-
LFOV	FOV,RESTRAINED BETWEEN 0 AND 1	-
LS	AMOUNT OF LIGNIN IN SHOOT	KG/HA
LSNHS	SINE HEIGHT OF YESTERDAY'S SUN	-
LTAIR	FLOW OF FRESH AIR, INJECTED INTO THE SYSTEM	LITER/S
LTAITB	TABLE OF FLOW OF FRESH AIR, INJECTED INTO THE SYSTEM	
LWC	LONG WAVE RADIATION PER LEAF AREA UNDER CLEAR SKY	J/M2(LEAF)/
LWO	LONG WAVE RADIATION PER LEAF AREA UNDER OVERCAST SKY	J/M2(LEAF)/
LWR	NET ABSORBED LONG WAVE RADIATION PER LEAF AREA	J/M2(LEAF)/
LWRCI	NET THERMAL RADIATION CLEAR SKY	J/M2/S
LWRI	ACTUAL NET INCOMING THERMAL RADIATION	J/M2/S
LWROI	NET THERMAL RADIATION OVERCAST SKY	J/M2/S
MAXT	MAXIMUM VALUE OF VARIABLE IN MACRO WAVE	-
MDPTTB	MINIMUM DAILY DEW POINT TEMPERATURE MEASURED VERSUS TIME	-
MINT	MINIMUM VALUE OF VARIABLE IN MACRO WAVE	-
MNTT	MINIMUM DAILY TEMPERATURE MEASURED VERSUS TIME	-
MXTT	MAXIMUM DAILY TEMPERATURE MEASURED VERSUS TIME	-
MR	AMOUNT OF MINERALS IN ROOTS	KG/HA
MRR	CO2 EVOLUTION RESULTING FROM MAINTENANCE OF THE ROOT	KG/HA/S
MRS	CO2 EVOLUTION RESULTING FROM MAINTENANCE OF THE SHOOT	KG/HA/S
MS	AMOUNT OF MINERALS IN SHOOT	KG/HA
N	RUNNER IN DO LOOP	-
NCAS	NET CO2-ASSIMILATION SHOOT	KG(CO2)/HA
NCASC	NET CO2-ASSIMILATION SHOOT CALCULATED	KG(CO2)/HA
NCASM	NET CO2-ASSIMILATION SHOOT MEASURED	KG(CO2)/HA
NCRIL	NET CO2-ASSIMILATION,INDIVIDUAL LEAVES	KG(CO2)/HA(LEAF)
NCRL	NET CO2-ASSIMILATION,ALL LEAVES IN CANOPY	KG(CO2)/HA
NCRLC	NET CO2-ASSIMILATION OF ALL LEAVES IN CANOPY UNDER CLEAR LIGHT CONDITIONS	KG(CO2)/HA
NCRLO	NET CO2-ASSIMILATION OF ALL LEAVES IN CANOPY UNDER OVERCAST LIGHT CONDITIONS	KG(CO2)/HA
NETR	NET RADIATION, ABSORBED BY CANOPY , PER GROUND AREA	J/M2/S
NETRS	NET RADIATION AT SOIL SURFACE	J/M2/S
NIR	ABSORBED NEAR-INFRARED RADIATION PER LEAF AREA	J/M2(LEAF)
NIRD	AUXILIARY VARIABLE	J/M2(LEAF)
NIRDFC	ABSORBED DIFFUSE NEAR-INFRARED RADIATION PER LEAF AREA UNDER A CLEAR SKY, EITHER IN THE TOP LAYER OF LAI=3 OR IN THE REST BELOW IT	J/M2(LEAF)
NIRDFO	SAME AS NIRDFC BUT UNDER AN OVERCAST SKY	J/M2(LEAF)
NIRT	AUXILIARY VARIABLE	J/M2(LEAF)
OAV	AVERAGE PROJECTION OF THE LEAVES IN THE DIFFERENT DIRECTIONS	-
OUTDEL	TIME INTERVAL FOR OUTPUTTING PLOT RESULTS	S
PAR	INTERMEDIATE VARIABLE	J/M2(LEAF)
PARL	CONSTANT IN FORMULA FOR CALCULATION OF BOUNDERY LAYER RESISTANCE AROUND THE LEAF	S**(0.5)
PI	CIRCUMFERENCE OF A CIRCLE , DIVIDED BY ITS DIAMETER	-
PR	AMOUNT OF PROTEIN IN ROOTS	KG/HA
PRDEL	TIME INTERVAL FOR OUTPUTTING PRINT RESULTS	S
PS	AMOUNT OF PROTEIN IN SHOOT	KG/HA
PSCH	PSYCHROMETRIC CONSTANT	MBAR/DEGREE
PROJ	RATIO OF THE AREA SHADED BY ENCLOSURE AND ITS ACTUAL GROUND AREA	-

NAME	DESCRIPTION	UNIT
Q10	INCREASE IN RATE OF MAINTENANCE PROCESSES PER 10 DEGREE C	-
RA	RESISTANCE OF BOUNDARY LAYER ROUND LEAF FOR HEAT	S/M
RAD	1 DEGREE IN RADIANS (180/PI)	-
RADCV	RADIATION CONVERSION FACTOR FOR DTRT INTO J/M2	-
RAF	RATE OF FORMATION OF ORGANIC ANIONS CONCURRENT WITH THE RATE OF NITRATE REDUCTION	KG/HA/S
RATE	INPUT VARIABLE IN MACRO	-
RCO2I	INTERNAL CONCENTRATION,MAINTAINED BY STOMATAL REGULATION	VPPM
RCO2IM	MAXIMUM INTERNAL CO2-CONCENTRATION	VPPM
RCRS	RELATIVE RATE OF CONSUMPTION OF RESERVES	1/S
RDPF	RELATIVE DIFFERENCE BETWEEN CARBON PRESENT IN PLANT AND CARBON IN NET FLUX INTO PLANT	-
REDFRL	REDUCTION FACTOR ACCOUNTING FOR FEEDBACK OF RESERVE LEVEL	-
REDFRT	REDFRL VERSUS RESERVE LEVEL(AS A FRACTION)	-
REFN	REFLECTION COEFFICIENT CANOPY WITH HORIZONTAL LEAVES NEAR-INFRARED	-
REFV	REFLECTION COEFFICIENT CANOPY WITH HORIZONTAL LEAVES IN VISIBLE REGION	-
RELPRO	SHADE OF ENCLOSURE ,AVERAGE OVER HEMISPHERE	-
RES	RESERVE CARBOHYDRATES (STARCH) IN PLANTS	KG/HA
RESCW	CUTICULAR RESISTANCE FOR TRANSPIRATION	S/M
RESL	RESERVE LEVEL IN PLANTS	-
RESPI	INITIAL FOR RESERVES PERCENTAGE	-
RFN	REFLECTION COEFFICIENT CANOPY FOR DIRECT RADIATION NEAR-INFRARED	-
RFOVN	REFLECTION COEFFICIENT CANOPY UNDER OVERCAST SKY FOR NEAR-INFRARED	-
RFOVV	REFLECTION COEFFICIENT CANOPY UNDER OVERCAST SKY FOR VISIBLE RADIATION	-
RFV	REFLECTANCE CANOPY FOR DIRECT RADIATION VISIBLE	-
RIECO2	RATIO INTERNAL/EXTERNAL CO2-CONCENTRATION	-
RH	RELATIVE HUMIDITY	-
RHOCP	VOLUMETRIC HEAT CAPACITY OF THE AIR	J/M3/DEGREE(C)
RISE	TIME OF SUNRISE	H
RISEI	INITIAL VALUE OF RISE	H
RMES	MESOPHYLL RESISTANCE FOR CO2 DIFFUSION	S/M
RLAI	EXCESS LEAF AREA INDEX ABOVE 3	M2/M2
RN	SEE RFN	-
RNO3	RATE OF NITRATE REDUCTION	KG(NO3)/HA/S
ROTC	TIME CONSTANT OF RATE OF ROTTING OLD ROOTS	S
RR	HEAT EXCHANGE RESISTANCE	DEGREE(C)*S*M2(LEAF)/J
RUPT	CO2 EVOLUTION RESULTING FROM THE UPTAKE OF MINERALS	KG(CO2)/HA/S
RV	SEE RFV	-
RWCLT	AUXILIARY VARIABLE TO CALCULATE RELATIVE WATER CONTENT PLANT	-
RWCP	RELATIVE WATER CONTENT PLANTS	-
RWCPL	AUXILIARY VARIABLE TO CALCULATE RELATIVE WATER CONTENT PLANT	-
S	DISTRIBUTION FUNCTION OF LEAVES IN 9 DIFFERENT INCLINATION CLASSES OVER SINES OF INCIDENCE FOR 9 INCLINATIONS OF THE SUN	-
SCN	SCATTERING COEFFICIENT OF LEAVES FOR NEAR-INFRARED	-
SCV	SCATTERING COEFFICIENT OF LEAVES FOR VISIBLE RADIATION	-
SELECT	DUMMY VARIABLE	-
SHL	SENSIBLE HEAT LOSS OF LEAVES PER LEAF AREA	J/M2(LEAF)/S
SHLA	SHADED LEAF AREA IN THE TOP PORTION OF LAI=3.	-
SI	INTERMEDIATE VARIABLE	-
SLLA	SUNLIT LEAF AREA	M2(LEAF)/M2(GROUND)
SLOPE	SLOPE OF SATURATED VAPOUR PRESSURE CURVE AT AIR TEMP.	MBAR/DEGREE(C)
SMU	STARCH REQUIRED FOR UPTAKE OF MINERALS AND NITRATE	KG/HA/S
SNDC	SINE DECLINATION OF SUN	-

NAME	DESCRIPTION	UNIT
SN	NUMBER OF CLASS OF SINE OF INCIDENCE	-
SNHS	SINE OF HEIGHT OF SUN BUT ZERO WHEN SUN BELOW HORIZON	-
SNHSS	SINE OF HEIGHT OF SUN, ALSO WHEN NEGATIVE	-
SNLT	SINE OF LATITUDE OF EXPERIMENTAL PLOT	-
SQ	INTERMEDIATE VARIABLES TO OBTAIN ARCSINE	-
SQNI	FACTOR OF DECREASE OF EXTINCTION COEFFICIENT FOR NEAR-INFRARED RADIATION	-
SQVI	FACTOR OF DECREASE OF EXTINCTION COEFFICIENT FOR VISIBLE RADIATION	-
SR	TOTAL STARCH REQUIREMENT OF THE ROOT	KG/HA/S
SRDOA	STARCH LOST BY CO2 EVOLUTION DURING DECARBOXYLATION OF ORGANIC ANIONS	KG/HA/S
SRES	LEAF RESISTANCE FOR TRANSPIRATION	S/M
SRESL	LEAF RESISTANCE FOR TRANSPIRATION,AS DETERMINED BY CO2 REGULATION	S/M
SRGCR	STARCH REQUIREMENT FOR CONVERSION AND TRANSLOCATION OF CARBOHYDRATES IN THE ROOT	KG/HA/S
SRGCS	STARCH REQUIREMENT FOR CONVERSION AND TRANSLOCATION OF CARBOHYDRATES IN THE SHOOT	KG/HA/S
SRGFR	STARCH REQUIREMENT FOR CONVERSION AND TRANSLOCATION OF FATS IN THE ROOT	KG/HA/S
SRGFS	STARCH REQUIREMENT FOR CONVERSION AND TRANSLOCATION OF FATS IN THE SHOOT	KG/HA/S
SRGLR	STARCH REQUIREMENT FOR CONVERSION AND TRANSLOCATION OF LIGNIN IN THE ROOT	KG/HA/S
SRGLS	STARCH REQUIREMENT FOR CONVERSION AND TRANSLOCATION OF LIGNIN IN THE SHOOT	KG/HA/S
SRGMR	STARCH REQUIREMENT FOR CONVERSION AND TRANSLOCATION OF MINERALS IN THE ROOT	KG/HA/S
SRGMS	STARCH REQUIREMENT FOR CONVERSION AND TRANSLOCATION OF MINERALS IN THE SHOOT	KG/HA/S
SRGPR	STARCH REQUIREMENT FOR CONVERSION AND TRANSLOCATION OF PROTEINS IN THE ROOT	KG/HA/S
SRGPS	STARCH REQUIREMENT FOR CONVERSION AND TRANSLOCATION OF PROTEINS IN THE SHOOT	KG/HA/S
SRGR	STARCH REQUIREMENT FOR GROWTH OF THE ROOT	KG/HA/S
SRGS	STARCH REQUIREMENT FOR GROWTH OF THE SHOOT	KG/HA/S
SRGSAR	STARCH REQUIREMENT FOR THE FORMATION OF SKELETONS OF AMINO ACIDS FOR ROOT PROTEINS	KG/HA/S
SRGSAS	STARCH REQUIREMENT FOR FORMATION OF SKELETONS OF AMINO ACIDS FOR SHOOT PROTEINS	KG/HA/S
SRMR	STARCH REQUIREMENT FOR MAINTENANCE OF THE ROOT	KG/HA/S
SRMS	STARCH REQUIREMENT FOR MAINTENANCE OF THE SHOOT	KG/HA/S
SRSOA	STARCH REQUIREMENT FOR FORMATION OF SKELETONS OF ORGANIC ANIONS	KG/HA/S
SRR	SHOOT-ROOT RATIO	-
SRTPAR	STARCH REQUIREMENT FOR TRANSPORT OF ORGANIC ANIONS IN ROOT	KG/HA/S
SRTPAS	STARCH REQUIREMENT FOR TRANSPORT OF ORGANIC ANIONS IN SHOOT	KG/HA/S
SRW	LEAF RESISTANCE FOR TRANSPIRATION,AS DETERMINED BY WATER POTENTIAL	S/M
SRCTB	STOMATAL CONDUCTANCE TABLE VERSUS RWCP	-
SS	TOTAL STARCH REQUIREMENT IN SHOOT	KG/HA/S
STDAY	DAY IN THE YEAR WHEN SIMULATION STARTS	DAY
SUBC	TIME CONSTANT OF SUBERIZATION OF YOUNG ROOTS	S
SUMF	SUM OF LEAF ANGLE DISTRIBUTION,SHOULD EQUAL 1	-
SUNDCL	DIRECT VISIBLE RADIATION STANDARD SKY CLEAR	J/M2/S
SUNDTB	DIRECT RADIATION FOR STANDARD SKIES TABLE	-
SUNPER	DIRECT SUN IRRADIATION, PERPENDICULAR ON THE BEAM EITHER FOR VISIBLE OR FOR NEAR-INFRARED(FIFTY/FIFTY DISTRIBUTION)	J/M2/S
SVP	SATURATED VAPOUR PRESSURE	MBAR
SYR	RATE OF SUBERIZATION OF YOUNG ROOTS	KG/HA/S

NAME	DESCRIPTION	UNIT
TA	AIR TEMPERATURE	DEGREE(C)
TANIRC	AUXILIARY VARIABLE FOR CALCULATING OF ANIRC	J/M2/S
TANIRO	AUXILIARY VARIABLE FOR CALCULATING OF ANIRO	J/M2/S
TATB	TABLE OF TEMPERATURE AS FUNCTION OF HOUR OF THE DAY	-
TAVISC	AUXILIARY VARIABLE FOR CALCULATING OF AVISC	J/M2/S
TAVISO	AUXILIARY VARIABLE FOR CALCULATING OF AVISO	J/M2/S
TC	TRANSPIRATION COEFFIENT(H2O TRANSPIRED/SHOOT FORMED(DRY))	G/G
TEFR	EFFECT OF TEMPERATURE ON RATE OF ROOT MAINTENANCE	-
TEFS	EFFECT OF TEMPERATURE ON RATE OF SHOOT MAINTENANCE	-
TEHL	EVAPORATIVE HEAT LOSS OF LEAVES PER GROUND AREA	J/M2/S
TEHLC	EVAPORATIVE HEAT LOSS PER GROUND AREA,SKY CLEAR	J/M2/S
TEHLO	EVAPORATIVE HEAT LOSS PER GROUND AREA,SKY OVERCAST	J/M2/S
TELLER	TELLER COUNTS HOW MANY TIMES PROGRAM IS UPDATED	-
TGTB	TABLE OF TEMPERATURE EFFECT ON GROWTH (MULTIPLICATION FACTOR)	-
TIM	USED IN MACRO TO CALCULATE WAVE	H
TIME	SIMULATED TIME ELAPSED SINCE START OF SIMULATION	S
TL	LEAF TEMPERATURE	DEGREE(C)
TNCAP	TOTALLED NET CO2-ASSIMILATION PLANTS	KG/HA
TNETR	TOTAL NET RADIATION	J/M2/S
TNETRC	TOTAL NET RADIATION UNDER CLEAR SKY	J/M2/S
TNETRO	TOTAL NET RADIATION UNDER OVERCAST SKY	J/M2/S
TOTDIF	ACCUMULATED DIFFERENCE BETWEEN TRANSPIRATION RATE AND RATE OF WATER UPTAKE IN ITERATION PROCEDURE	G/M2
TR	ACCUMULATED WHOLE PLANT DISSIMILATION	KG(CO2)/HA
TRC	TRANSPIRATION RATE OF CANOPY	G/M2/S
TRCCH	TRANSPIRATION RATE OF CANOPY CALCULATED	G(H2O)/M2/H
TRCMH	TRANSPIRATION RATE OF CANOPY MEASURED	G(H2O)/M2/H
TRCMTB	TABLE OF MEASURED RATE OF TRANSPIRATION OF FIELD CROP IN TIME	-
TRPH	NAME OF MACRO,DESCRIBING ENERGY BALANCE OF INDIVIDUAL LEAVES	-
TRRT	RATE OF TRANSPORT OF ORGANIC ANIONS TO THE ROOT FOR DECARBOXYLATION	KG/HA/S
TS	SOIL TEMPERATURE IN ROOT ZONE	DEGREE(C)
TSHL	SENSIBLE HEAT LOSS OF LEAVES PER GROUND AREA	J/M2/S
TSHLC	SENSIBLE HEAT LOSS PER GROUND AREA,SKY CLEAR	J/M2/S
TSHLO	SENSIBLE HEAT LOSS PER GROUND AREA,SKY OVERCAST	J/M2/S
TTRCC	TOTALLED TRANSPIRATION OF CANOPY CALCULATED	G(H2O)/M2
TTRCM	TOTALLED TRANSPIRATION OF CANOPY MEASURED	G(H2O)/M2
TWS	TOTAL WEIGHT OF SHOOT AND RESERVES	KG/HA
TWT	TOTAL WEIGHT CROP	KG/HA
TWUR	TOTALLED WATER UPTAKE BY ROOT SYSTEM	G(H2O)/M2(GROUND)
URES	CONSUMPTION RATE OF UTILIZATION OF RESERVES	KG(STARCH)/HA/S
VAL	VALUE — USED IN MACRO TO CALCULATE WAVE	-
VALAMP	AMPLITUDE — USED IN MACRO TO CALCULATE WAVE	-
VALAV	AVERAGE VALUE — USED IN MACRO TO CALCULATE WAVE	-
VALSS	VALUE AT SUNSET — USED IN MACRO TO CALCULATE WAVE	-
VALSR	VALUE AT SUNRISE — USED IN MACRO TO CALCULATE WAVE	-
VAPHT	HEAT OF VAPORIZATION OF WATER	J/G
VIS	ABSORBED VISIBLE RADIATION PER LEAF AREA	J/M2(LEAF)/S
VISD	AUXILIARY VARIABLE	J/M2(LEAF)/S
VIST	AUXILIARY VARIABLE	J/M2(LEAF)/S
VISDFC	SAME AS NIRDFC BUT FOR VISIBLE RADIATION UNDER A CLEAR SKY	J/M2(LEAF)/S
VISDFO	SAME AS NIRDFO BUT FOR VISIBLE RADIATION	J/M2(LEAF)/S
VPMI	CO2 CONCENTRATION OF AIR FLOWING INTO THE ENCLOSURE	VPPM
VPMITB	TABLE OF VPMI VERSUS TIME	-
VPMOC	CALCULATED CO2-CONCENTRATION OF AIR FLOWING OUT OF ENCLOSURE	VPPM
VPMOM	MEASURED CO2-CONCENTRATION OF AIR FLOWING OUT OF ENCLOSURE	VPPM
VPMOTB	TABLE OF VPMOM VERSUS TIME	-

NAME	DESCRIPTION	UNIT
WCF	WEIGHT OF ACCUMULATED CARBON OF THE NET FLUX INTO PLANTS	KG(C)/HA
WCP	WEIGHT OF CARBON IN PLANTS	KG(C)/HA
WCPL	WATER CONTENT PLANTS	KG(H2O)/HA
WCRR	WEIGHT CONDUCTANCE RATIO OF ROOT SYSTEM	(KG/HA)/(G(H2O)/BAR/S))
WDL	AVERAGE WIDTH OF THE LEAVES	M
WGRTB	RELATION BETWEEN RELATIVE WATER CONTENT AND GROWTH OF ROOT	-
WGSTB	RELATION BETWEEN RELATIVE WATER CONTENT AND GROWTH OF SHOOT	-
WOR	WEIGHT OF OLD ROOTS	KG/HA
WPTC	WATER POTENTIAL OF PLANTS	BAR
WPTSL	WATER POTENTIAL OF SOIL	BAR
WPTTB	TABLE PLANT WATER POTENTIAL VS PLANT RELATIVE WATER CONTENT	-
WRC	WEIGHT ROOT SYSTEM CALCULATED	KG/HA
WS	WIND SPEED IN THE ENCLOSURE	M/S
WSC	WEIGHT SHOOT CALCULATED	KG/HA
WSM	WEIGHT SHOOT MEASURED	KG/HA
WSN	WEIGHT SHOOT MEASURED AT DAY N + 1	KG/HA
WSO	WEIGHT SHOOT MEASURED AT DAY N	KG/HA
WSMTB	TABLE OF MEASURED SHOOT WEIGHT IN TIME	-
WSTB	TABLE OF WIND SPEED IN THE ENCLOSURE AS FUNCTION OF HOUR	-
WUR	RATE OF WATER UPTAKE BY ROOT SYSTEM	G/M2/S
WYR	WEIGHT OF YOUNG ROOTS	KG/HA
Z	DISTRIBUTION FUNCTION OF ALL LEAVES TOGETHER OVER CLASSES OF SINES OF INCIDENCE	-
ZA	CUMULATIVE FUNCTION OF Z	-
ZISSN	SAME AS Z	-
ZHOLD	CSMP FUNCTION	-

Index

Books already published in this series:

C.T. de Wit and H. van Keulen. Simulation of transport processes in soils (2nd ed., revised). 1975. 109 pp. ISBN 90-220-0591-7. Price Dfl. 15.00

J. Beek and M.J. Frissel. Simulation of nitrogen behaviour in soils. 1973. 76 pp. ISBN 90-220-0440-6. Price Dfl. 12.50

H. G. Fransz. The functional response to prey density in an acarine system. 1974. 149 pp. ISBN 90-220-0509-7. Price Dfl. 20.00

M.J. Frissel and P. Reiniger. Simulation of accumulation and leaching in soils. 1974. 124 pp. ISBN 90-220-0530-5. Price Dfl. 17.50

H. van Keulen. Simulation of water use and herbage growth in arid regions. 1975. 184 pp. ISBN 90-220-0557-7. Price Dfl. 27.50

G.F. Makkink and H.D.J. van Heemst. Simulation of the water balance of arable land and pastures. 1975. 87 pp. ISBN 90-220-0566-6. Price Dfl. 15.00

G.W. Arnold and C.T. de Wit (Eds). Critical evaluation of systems analysis in ecosystems research and management. 1976. 114 pp. ISBN 90-220-0593-3. Price Dfl. 20.00

J. van den Bos and R. Rabbinge. Simulation of the fluctuations of the grey larch bud moth. 1976. 91 pp. ISBN 90-220-0589-5. Price Dfl. 16.50

R. Rabbinge. Biological control of fruit-tree red spider mite. 1976. 234 pp. ISBN 90-220-590-8. Price Dfl. 37.50

J. Goudriaan. Crop micrometeorology: a simulation study. 1977. 257 pp. ISBN 90-220-0614-X, Price Dfl. 37.50

E. van Elderen. Heuristic strategy for scheduling farm operations. 1977. 233 pp. ISBN 90-220-0612-3. Price Dfl. 22.50

P. Kampmeijer and J.C. Zadoks. EPIMUL, a simulator of foci and epidemics in mixtures of resistant and susceptible plants, mosaics and multilines. 1977. 56 pages. ISBN 90-220-0636-0. Price Dfl. 10.00

T. Kozai, J. Goudriaan and M. Kimura. Light transmission and photosynthesis in greenhouses. 1978. 105 pp. ISBN 90-220-0646-8. Price Dfl. 16.50

K.R. Chistian et al. Simulation of grazing systems. 1978. 121 pp. ISBN 90-220-0645-X. Price Dfl. 18.00

C.T. de Wit and J. Goudriaan. Simulation of ecological processes. (2nd ed. revised and extended). 1978. 183 pp. ISBN 90-220-0652-2. Price Dfl. 22.50 (Published by Pudoc outside the USA, Canada and Latin America)